CO-ORDINATE GEOMETRY

DPH MATHEMATICS SERIES

CO-ORDINATE GEOMETRY

By

A.K. Sharma

DISCOVERY PUBLISHING HOUSE
NEW DELHI-110002

Reprinted - 2019

First Published - 2005

ISBN: 978-81-8356-028-3

Co-ordinate Geometry 2-D

Published by:

DISCOVERY PUBLISHING HOUSE PVT. LTD.
4383/4B, Ansari Road, Darya Ganj
New Delhi-110 002 (India)
Phone: +91-11-23279245, 23253475; 43596065
E-mail: discoverybooksindia@gmail.com
discoverypublishinghouse@gmail.com
web: www.discoverypublishinggroup.com

Printed at:
Infinity Imaging Systems
Delhi

Preface

This book of Co-ordinate Geometry has been especially written to meet the requirements of B.A/B.Sc. students of all Indian Universities.

The subject matter has been discussed in such a simple way that the student will find no difficulty to understand it. The proof of various theorems and examples have been given with minute details each chapter of this books. Contains complete theory and large number of solved examples sufficient problems have also been selected from various Indian Universities and competitive examination.

I hope that this book warmly received by the student and teachers.

I wish to thank my colleague and friends to countless helpful remarks and suggestions. The author wish to express his thanks to the publisher M/s Discovery Publishing House, New Delhi for bringing out this book in present nice form.

A.K. Sharma

Preface

This book of Co-ordinate Geometry has been especially written to meet the requirements of B.A. / B.Sc. Students of all Indian Universities.

The subject matter [illegible] that the [illegible] may [illegible].

[illegible] solved [illegible].

I hope that this book will be received by the students and teachers.

I will be thankful to the readers and the teachers for their valuable remarks and suggestions. The author will be pleased to receive [illegible] for improving the book in the future edition.

A.K. Sharma

Contents

Contents

Page

1

INTRODUCTION TO CO-ORDINATE GEOMETRY

INTRODUCTION

French philoshopher and mathematician Rene Descartes (1596-1665) published "La-Geomatric" in 1637 which he induced the analytical approach by systematically using algebra in his study of geometry. This was achieved by representing point in the plane by ordered pari of real numbers called Cartesian Coordinates name after Rene (Descartes) and representing lines and curve by algebraic equations. This wedding of algebra and Geometry is known as analytic or coordinate geometry.

CO-ORDINATES OF A POINT: RECTANGULAR AXES

Let XOX' and YOY' be two perpendicular lines in the Euclidean plane intersecting each other at a point.

(i) The line XOX' is called the *x-axis.*

(ii) The line YOY' is called the *y-axis.*

Both these lines taken in this very order are called the *rectangular axes* or the *axes of co-ordinates* or simply the *axes.*

(iii) Their point of intersection O is known the *origin.*

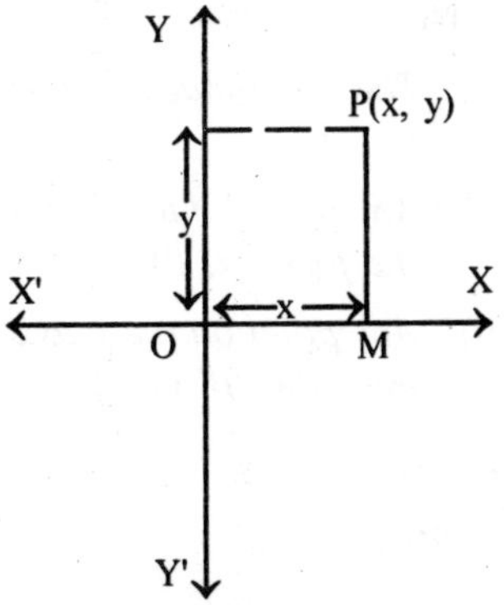

Let P be any point in the plane. From P draw PM perpendicular to XOX'. The lengths OM and PM determine the position of the point.

Now let OM = x and MP = y. The numbers x and y together are called the *Cartesian co-ordinates* of P and are denoted by P(x, y).

(i) the length OM = x is called the *x-coordinate* or *abscissa* of P.

(ii) the length MP = y is called the *y-coordinate* or *ordinate* of P.

The symbol P(x, y) is used to denote the point P in which *the abscissa is always written first and separated from the ordinate by a comma.*

QUADRANT AND THEIR SIGN

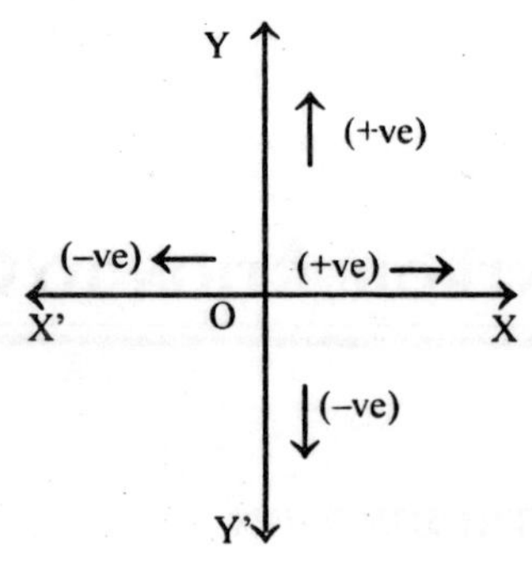

(i) For distances along the x-axis, positive values are measured to the right of the origin ande negative values to the left of origin.

(ii) For distances along the y-axis, positivge values are measured upward and negative values are measured upward and negative values are measured downward.

In accordance with the above convention, the signs of the coordinates of a point in different quadrants are as follows:

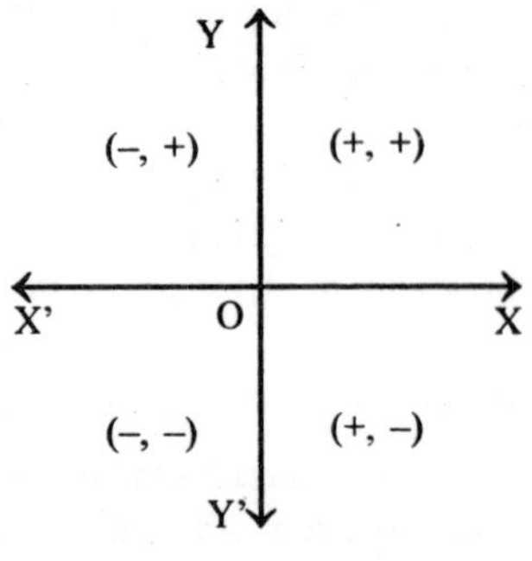

1st quadreant	x = +ve, y = +ve
2nd quadreant	x = –ve, y = +ve
3rd quadreant	x = –ve, y = –ve
4th quadreant	x = +ve, y = –ve

Note:

(i) *The coordinates of the origin are (0, 0).*

(ii) *Any point on the x-axis is of the form (x, 0).*

(iii) *Any point on the y-axis is of the form (0, y).*

Example:

Illustration.

Plot the points

(i) (2, 5);

(ii) (–2, 5);

(iii) (–3, –3);

(iv) (4, –6).

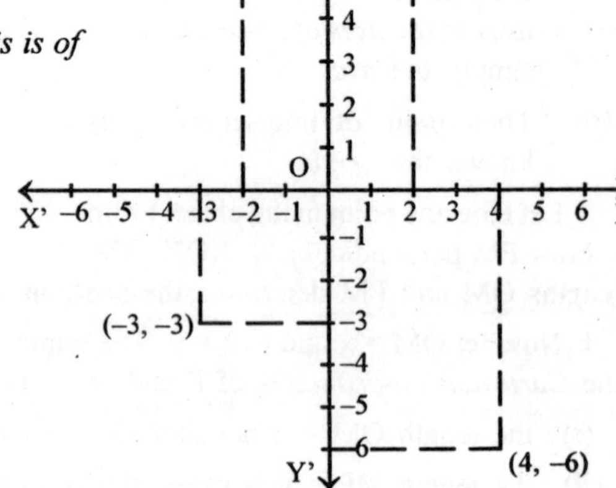

Solution:

Let XOX' and YOY' be the co-ordinate axes. These four points will be plotted as shown in figure.

DISTANCE FORMULA : DISTANCE BETWEEN TWO POINTS

The distance between two points (x_1, y_1) and (x_2, y_2) is

$$d = \sqrt{(x_2 - x_1) + (y_2 - y_1)^2}$$

Proof:

Let $P(x_1, y_1)$ and $Q(x_2, y_2)$ be two given points. From P and Q draw perpendiculars PM and QN to OX respectively. Again from P draw PR ⊥ NQ.

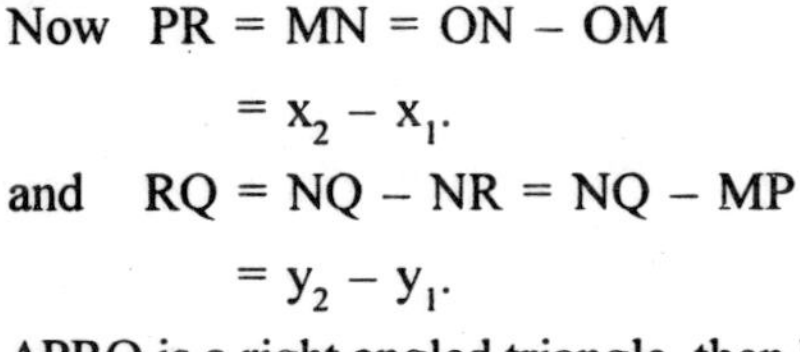

Now $PR = MN = ON - OM$

$= x_2 - x_1.$

and $RQ = NQ - NR = NQ - MP$

$= y_2 - y_1.$

ΔPRQ is a right angled triangle, then by Pythagoras theorem,

we have $PQ^2 = PR^2 + RQ^2$

$= (x_2 - x_1)^2 + (y_2 - y_1)^2.$

$$\therefore \quad PQ = \sqrt{(x_2 - x_1)^2 + (y_2 - y_1)^2}.$$

Cor : Distance from the origin

The distance of the point $P(x_1, y_1)$ from the origin (0, 0) is

$$\sqrt{(x_1 - 0)^2 + (y_1 - 0)^2}.$$

or $\quad = \sqrt{x_1^2 + y_1^2}.$

Note. *Distance between two given points*

$$= \sqrt{(\text{Difference of Abscissae})^2 + (\text{Difference of Ordinates})^2}$$

Example 1:

Prove that the quadrilateral with the vertices (2, –1), (3, 4), (–2, 3) and (–3, –2) is a rhombus.

Solution:

Let (2, –1), (3, 4), (–2, 3) and (–3, –2) be denoted by the point A, B, C and D, respectively.

$$\therefore \quad AB = \sqrt{(2-3)^2 + (-1-4)^2} = \sqrt{26},$$

$$BC = \sqrt{[3-(-2)]^2 + (4-3)^2} = \sqrt{26},$$

$$CD = \sqrt{[(-2)-(-3)]^2 + [3-(-2)]^2} = \sqrt{26},$$

and
$$DA = \sqrt{[(-3)-(2)]^2 + [(-2)-(-1)]^2} = \sqrt{26}.$$

Thus AB = BC = CD = AD.

Hence, ABCD is a rhombus.

Example 1:

Prove that the points (–2, 2) (8, –2) and (–4, –3) are the vertices of a right-angled triangle.

Solution:

Let the points (–2, 2), (8, –2) and (–4, –3) represent the point A, B and C respectively. Then

$$AB = \sqrt{[(8-(-2)]^2 + (-2-2)^2}$$

$$= \sqrt{10^2 + (-4)^2} = \sqrt{100+16} = \sqrt{116}.$$

$$BC = \sqrt{(-4-8)^2 + (-3+2)^2}$$

$$= \sqrt{12^2 + 1^2} = \sqrt{144+1} = \sqrt{145}.$$

$$CA = \sqrt{(-2+4)^2 + (2+3)^2}$$

$$= \sqrt{2^2 + 5^2} = \sqrt{4+25} = \sqrt{29}.$$

Thus,

$AB^2 = 116$, $BC^2 = 145$

and $CA^2 = 29$.

$\therefore AB^2 + CA^2 = 116 + 29 = 145 = BC^2$.

Hence ΔABC is a right-angled triangle.

SECTION FORMULAE

Given two points A (x_1, y_1) and B (x_2, y_2) the coordinates of the point P on AB which divides the line in the ratio lim (internally) are given by

$$x = \frac{mx_1 + lx_2}{m + l} = \frac{my_1 + ly_2}{l + m}$$

Proof:

Let the $A(x_1, y_1)$ and $B(x_2, y_2)$ be two given points and P be a point on AB which divides it in the given ration m : n. We have to find the coordinates of P. Let they are (x, y). Draw the perpendiculars AL, PM, BN on OX and AK, PT on PM and BN respectively. Then from similar triangles ΔAPK and ΔPBT, we know that

$$\frac{AP}{PB} = \frac{AK}{PT} = \frac{KP}{TB} \qquad \text{...(i)}$$

We have AP : PB = m : n,

$\therefore AK = LM = OM - OL = x - x_1$,

$PT = MN = ON - OM = x_2 - x$,

$KP = MP - MK = MP - LA = y - y_1$,

$TB = NB - NT = NB - MP = y_2 - y$,

$\therefore$ From (i), we have

$$\frac{m}{n} = \frac{x - x_1}{x_2 - x} = \frac{y - y_1}{y_2 - y}.$$

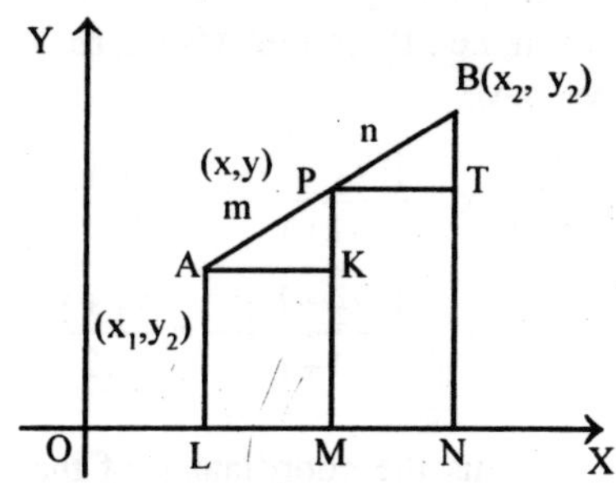

The first two relations give

$$\frac{m}{n} = \frac{x - x_1}{x_2 - x}$$

or $mx_2 - mx = nx - nx_1$

or $x(m + n) = mx_2 + nx_1$

or $x = \dfrac{mx_2 + nx_1}{m + n}$

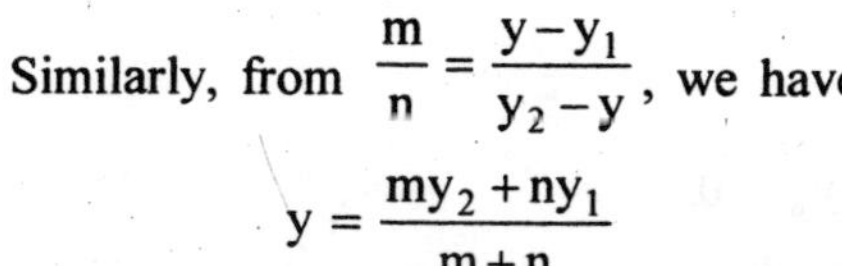

Similarly, from $\dfrac{m}{n} = \dfrac{y - y_1}{y_2 - y}$, we have

$$y = \frac{my_2 + ny_1}{m + n}$$

Thus, the co-ordinates of P are

$$x = \frac{mx_2 + nx_1}{m+n},\ y = \frac{my_2 + ny_1}{m+n}.$$

Note. *The above result can be remembered with the help of the following diagram.*

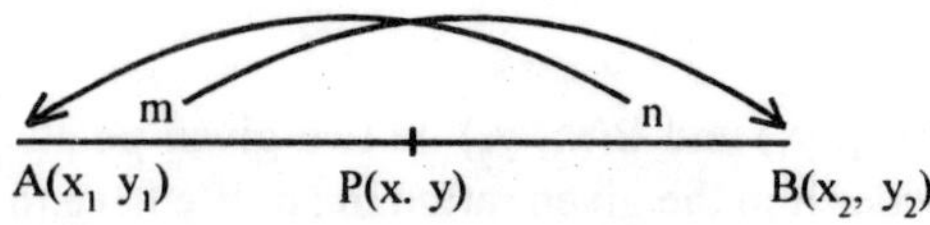

$$x = \frac{m(\text{abscissa of B}) + n(\text{abscissa of A})}{m+n = (\text{sum of the ratios})}$$

$$y = \frac{m(\text{ordinate of B}) + n(\text{ordinate of A})}{m+n = (\text{sum of the ratios})}$$

Cor : Mid Point Formula.

To find the co-ordinates of the midle point of the line joining two given points (x_1, y_1). and (x_2, y_2)

Proof:

Let P(x, y) be the mid-point of the join of $A(x_1, y_1)$ and $B(x_2, y_2)$, then m = n, i.e., P divides AB internally in the ratio 1 : 1. Then by section formula we have

$$\therefore\ x = \frac{1.x_2 + 1.x_1}{1+1} = \frac{x_1 + x_2}{2},$$

$$y = \frac{1.y_2 + 1.y_1}{1+1} = \frac{y_1 + y_2}{2}.$$

Thus the coordinates of the mid-point are $P\left(\frac{x_1 + x_2}{2} = \frac{y_1 + y_2}{2}\right)$.

Example:

Find the point at which the join of (2, 0) and (–3, 5) is divided in the ratio of 2 : 3 internally.

Solution:

Here $x_1 = 2,\ x_2 = -3,\ y_1 = 0,$

$y_2 = 5,\ m = 2,\ n = 3.$

Consider (x, y) as the required point then

$$x = \frac{mx_2 + nx_1}{m+n} = \frac{2(-3) + 3.2}{2+3} = \frac{0}{5} = 0.$$

Similarly, $$y = \frac{2\times5+3\times0}{2+3} = \frac{10+0}{5} = 2.$$

Thus the co-ordinates of the required point are (0, 2).

FORMULA FOR EXTERNAL DIVISION

To find the co-ordinates of the point which divide the straight line joining two given points externally in the ration m : n.

Proof:

Let $A(x_1, y_1)$ and $B(x_2, y_2)$ be two given points. Suppose P(x, y) is the point on AB which divides it externally on the given ratio m : n.

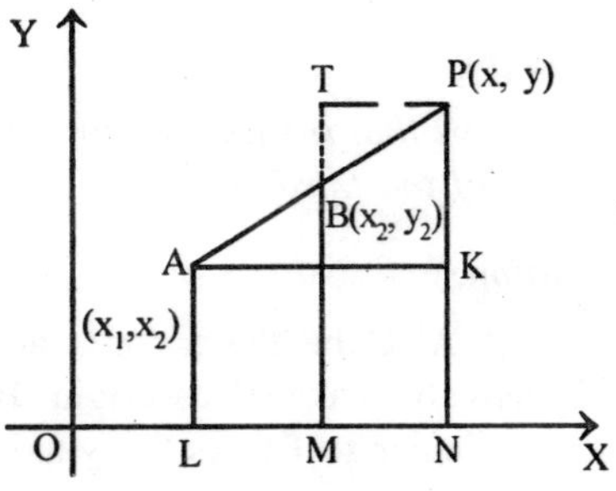

From A, B and P draw perpendicular AL, BM and PN ⊥ s on the x-axis. Through P draw a line parallel to OX meeting MB produced in T. From A draw AK ⊥ NP. Then from the similar Δs APK and ΔPBT, we have

$$\frac{AP}{PB} = \frac{AK}{PT} = \frac{KP}{TB}.$$

Now we have AP : PB = m : n and

$AK = LN = ON - OL = x - x_1,$

$PT = NM = ON - OM = x - x_2,$

$KP = PN - KN = y - y_1,$

$TB = TM - BM = PN - BM = y - y_2.$

Thus $$\frac{m}{n} = \frac{x - x_1}{x - x_2} = \frac{y - y_1}{y - y_2}$$

Now $$\frac{m}{n} = \frac{x \quad x_1}{x - x_2} \text{ given}$$

$$x = \frac{mx_2 - nx_1}{m-n}$$

and $$\frac{m}{n} = \frac{y - y_1}{y - y_2}$$ gives

$$y = \frac{my_2 - ny_1}{m - n}$$

Hence, the co-ordinates of P for external division are

$$x = \frac{mx_2 - nx_1}{m - n},\ y = \frac{my_2 - ny_1}{m - n}.$$

Note. *To obtain the cordinates of the point of external division, change n to –n and write down the co-ordinates of the point which divides the join of the given points internally in the ratio m : –n.*

Example:

Prove that the line joining the mid point of any two sides of a triangle is half of the third side.

Solution:

Let ABC be the triangle and D, E be the mide points of AB, AC respectively. Take. B as origin. BC as the axis of x and a line through B perpendicular to BC as the y-axis.

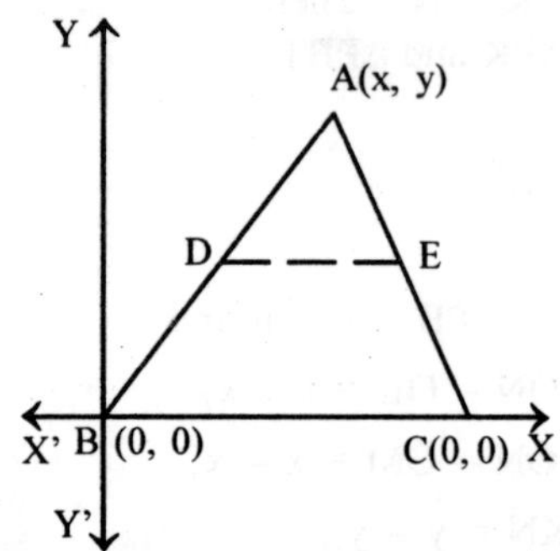

Let BC = a,

so that the point C is (a, 0) and let the vertex A be (x, y). The coordinates of the points D and E are respectively

$$\left(\frac{x}{2}, \frac{y}{2}\right), \text{ and } \left(\frac{x+a}{2}, \frac{y}{2}\right).$$

$$\therefore\ DE^2 = \left(\frac{x+a}{2} - \frac{x}{2}\right)^2 + \left(\frac{y}{2} - \frac{y}{2}\right)^2 = \left(\frac{a}{2}\right)^2$$

$$\Rightarrow \qquad DE = \frac{a}{2} \Rightarrow DE = \frac{1}{2} BC.$$

CO-ORDINATES OF THE CENTROID

To find the co-ordinates of the centroid of a triangle whose vertices are given.

Proof:

We know that the medians of a triangle are concurrent and the point at which they meet is called the centroid. The centroid divides each median in the ratio 2 : 1.

Let $A(x_1, y_1)$, $B(x_2, y_2)$, $A(x_3, y_3)$ be the given vertices of ΔABC. Let AP, BQ and CR be the medians and they intersect at the point G. Thus we have

$$\frac{AG}{GP} = \frac{2}{1}$$

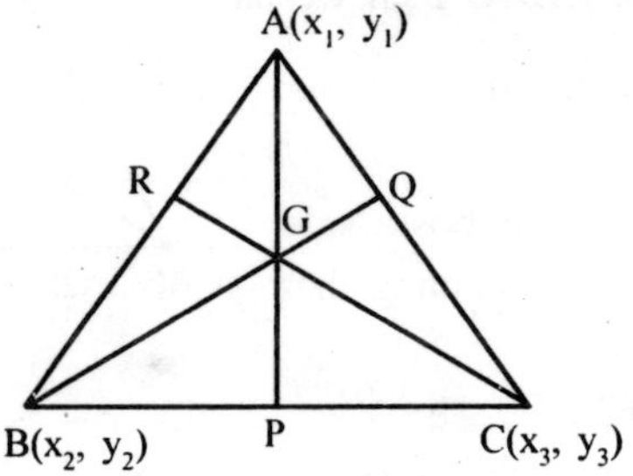

Now P is the mid-point of BC, then the co-ordinates of P are by

$$\left(\frac{x_2 + x_3}{2}, \frac{y_2 + y_3}{2}\right)$$

The co-ordinates of A are (x_1, y_1) and let the co-ordinates of G be (x, y), then we have

$$x = \frac{2.\frac{x_2 + x_3}{2} + 1.x_1}{2+1} = \frac{x_1 + x_2 + x_3}{3}$$

and $$y = \frac{2.\frac{y_2 + y_3}{2} + 1.y_1}{2+1} = \frac{y_1 + y_2 + y_3}{3}$$

Hence, the co-ordinates of the centroid are given

$$\left(\frac{x_1 + x_2 + x_3}{3}, \frac{y_1 + y_2 + y_3}{3}\right).$$

CO-ORDINATES OF THE INCENTRE

To find the co-ordinates of the incentre of a triangle whose vertices are given.

Proof:

We know tht the internal bisectors of the angles of a trinagle are concurrent and the point where they meet is known as the *incentre* of the triangle

Now let $A(x_1, y_1)$, $B(x_2, y_2)$ and $C(x_3, y_3)$ be the given vertices of the ΔABC, and let the lengths of sides be a, b and c respectively.

Again let AD be the bisector of the $\angle BAC$ and BI be the bisector of the $\angle ABC$, meeting AD at I, then I is the incentre.

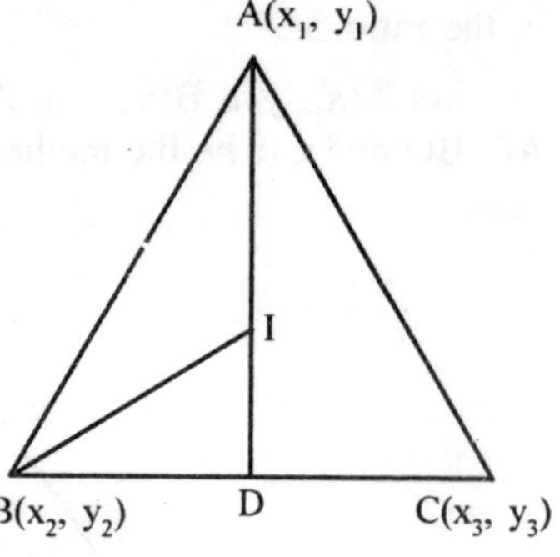

Now AD bisects $\angle BAC$ then we have

$$\Rightarrow \quad \frac{BD}{DC} = \frac{AB}{AC} = \frac{c}{b}.$$

($\because$ bisector divides the opposite side in the ratio of the sides containing that angle.)

$$\therefore \quad BD : DC :: c : b$$

$\Rightarrow$ co-ordinates of the point D are given by

$$\left(\frac{cx_3 + bx_3}{b+c}, \frac{cy_3 + by_3}{b+c}\right).$$

Agin we have $\frac{DC}{BD} = \frac{b}{c}$

$$\Rightarrow \quad 1 + \frac{DC}{BD} = 1 + \frac{b}{c} \quad \text{\{adding (i) both sides\}}$$

$$\Rightarrow \quad \frac{BD + DC}{BD} = \frac{c+b}{c}$$

$$\Rightarrow \quad \frac{BC}{BD} = \frac{c+b}{c} \Rightarrow \frac{BD}{BC} = \frac{c}{c+b}$$

$$\Rightarrow \quad \frac{BD}{a} = \frac{c}{c+b} \Rightarrow BD = \frac{ac}{b+c}. \quad [\because BC = a]$$

Again $\dfrac{AI}{ID} = \dfrac{AB}{BD}$ $\quad [\because \text{BI bisects } \angle ABD]$

$$= \frac{c}{\dfrac{ac}{b+c}} = \frac{b+c}{a}.$$

i.e. I divides AD in the ratio of b + c : a.

Now, if (x, y) are the co-ordinates of I, then

$$x = \frac{(b+c)\times\dfrac{cx_3+bx_2}{b+c}+ax_1}{b+c+a} = \frac{ax_2+bx_2+cx_3}{a+b+c}$$

and $$y = \frac{(b+c)\times\dfrac{cy_3+by_2}{b+c}+ay_1}{a+b+c} = \frac{ay_1+by_2+cy_3}{a+b+c}.$$

Hence, the co-ordinates of the incentre are

$$\left(\frac{ax_1+bx_2+cx_3}{a+b+c}, \frac{ay_1+by_2+cy_3}{a+b+c}\right).$$

Example

Find the incentre of a triangle whose vertices are (0, 0), (–3, 4) and (8, 6).

Solution:

Let A(0, 0), B(–3, 4) and C(8, 6) be the vertices of the triangle and let BC = a, CA = b and AB = c, then

$$a = \sqrt{(8+3)^2+(6-4)^2} = 5\sqrt{5}$$

$$c = \sqrt{(-3-0)^2+(4-0)^2} = 5$$

and $$b = \sqrt{(8-0)^2+(6-0)^2} = 10.$$

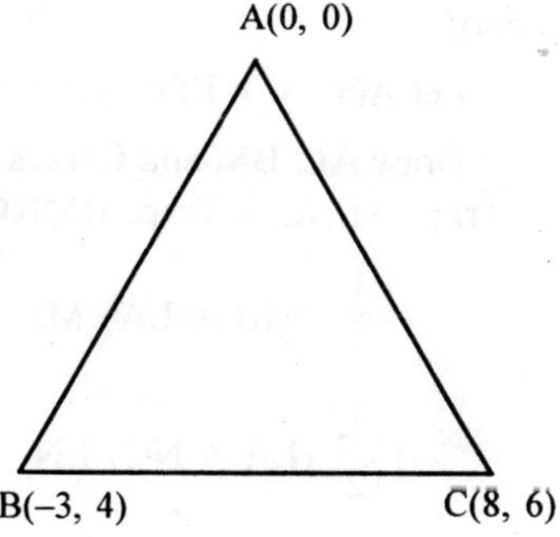

The co-ordinates of the incentre are

$$x = \frac{ax_1+bx_2+cx_3}{a+b+c}$$

$$= \frac{5\sqrt{5}\times 0+10\times(-3)+5\times 8}{5\sqrt{5}+10+5}$$

$$= \frac{10}{5\left(3+\sqrt{5}\right)} = \frac{2\left(3-\sqrt{5}\right)}{\left(3+\sqrt{5}\right)\left(3-\sqrt{5}\right)}$$

$$= \frac{2\left(3-\sqrt{5}\right)}{9--5} = \frac{3-\sqrt{5}}{2}.$$

$$y = \frac{ay_1+by_2+cy_3}{a+b+c} = \frac{5\sqrt{5\times 0+10\times 4+5\times 6}}{5\sqrt{5+10+5}}$$

$$= \frac{70}{5\left(\sqrt{3}+\sqrt{5}\right)} = \frac{14\left(3-\sqrt{5}\right)}{\left(3+\sqrt{5}\right)\left(3-\sqrt{5}\right)}$$

$$= \frac{14\left(3-\sqrt{5}\right)}{4} = \frac{7}{2}\left(3-\sqrt{5}\right).$$

Hence, the co-ordinates of the incentre are

$$\left[\frac{3-\sqrt{5}}{2}, \frac{7\left(3-\sqrt{5}\right)}{2}\right].$$

AREA OF A TRIANGLE

TO FIND AREA OF A TRIANGLE WHOSE VERTICES ARE GIVEN

Proof:

Let $A(x_1, y_1)$, $B(x_2, y_2)$ and $C(x_3, y_3)$ be the vertices of a triangle ABC.

Draw AL, BM and CN ⊥s on the x-axis. Then as ΔABC = Trap. ABML + Trap. ALNC – Trap. BMNC

$$= \frac{1}{2}\text{ (MB + LA) ML}$$

$$+ \frac{1}{2}\text{ (LA + NC) LN}$$

$$- \frac{1}{2}\text{ (MB + NC) MN}$$

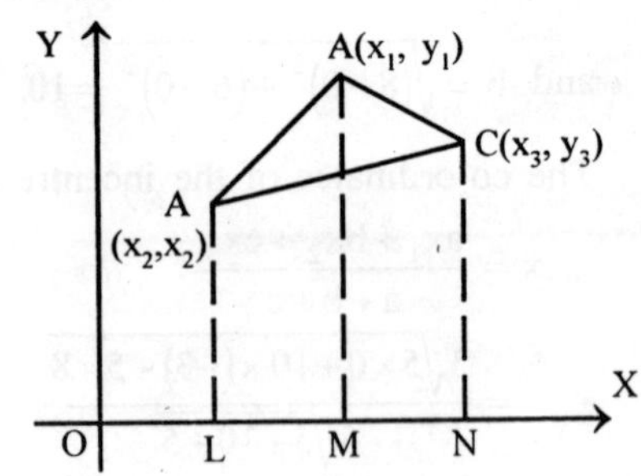

$$= \frac{1}{2}[(y_2 + y_1)(OL - OM) + (y_1 + y_3)(ON - OL) - (y_2 + y_3)(NL - OM)]$$

$$= \frac{1}{2}(y_2 + y_1)(x_1 - x_2) + (y_1 + y_3)(x_3 - x_1) - (y_2 + y_3)(x_3 - x_2)$$

$$= \frac{1}{2}[x_1 y_2 + x_1 y_1 - x_2 y_2 - x_2 y_1 + x_3 y_1 + x_3 y_3 - x_1 y_1 - x_1 y_3 - (x_3 y_2 + x_3 y_3 - x_2 y_2 - x_2 y_3)]$$

$$= \frac{1}{2}[(x_1 y_2 - x_2 y_1) + (x_2 y_3 - x_3 y_2) + (x_3 y_1 - x_1 y_3)]$$

$\therefore$ Area of ΔABC

$$= \frac{1}{2}[(x_1 y_2 - x_2 y_1) + (x_2 y_3 - x_3 y_2) + (x_3 y_1 - x_1 y_3)]$$

$$\text{Area of } \Delta ABC = \frac{1}{2}\ x_1(y_2 - y_3) + x_2(y_3 - y_1) + x_3(y_1 - y_2)$$

$$= \frac{1}{2}\begin{vmatrix} x_1 & y_1 & 1 \\ x_2 & y_2 & 1 \\ x_3 & y_3 & 1 \end{vmatrix}$$

Cor. 1. *Area of a quadrilaterial can be found by divided it into two triangles. In any quadrilateral ABCD if we draw diagonal AC, then quadrilateral*

$$ABCD = \Delta ABC + \Delta ACD$$

Cor. 2. *If the area of triangle obtained from the three given points is zero; the three points lie on a line. Therefore, conditions for three points to be collinear is :*

$$\frac{1}{2}[x_1(y_2 - y_3) + x_2(y_3 - y_1) + x_3(y_1 - y_2)] = 0$$

$$\Rightarrow \quad x_1(y_2 - y_3) + x_2(y_3 - y_1) + x_3(y_1 - y_2)] = 0.$$

Example:

A, B are the two points (3, 4), and (5, –2); find the point P such that PA = PB and the area of ΔPAB = 10.

Solution:

Let the coordinates of P be (x, y)

$\because$ PA = PB $\quad \therefore \quad PA^2 = PB^2$.

$\therefore (x - 3)^2 = (y - 4)^2 = (x - 5)^2 + (y + 2)^2$

or $x^2 - 6x + 9 + y^2 - 8y + 16 = x^2 - 10x + 25 + y^2 + 4y + 4$

i.e., $x - 3y - 1 = 0.$ ...(i)

Also $\Delta PAB = 10.$

But $\Delta PAB = \frac{1}{2} [4x - 3y - 6 - 20 + 5y + 2x]$

$= \pm (3x + y - 13).$ ...(ii)

From (ii), $3x + y - 13 = \pm 10$

i.e., $3x + y - 23 = 0$ ...(iii)

or $3x + y - 3 = 0.$ ...(iv)

Solving (i) and (iii), we get

$10x - 70 = 0.$

$\therefore$ $x = 7$ and $y = 2$

Again solving (i) and (iv), we get

$10x - 10 = 0.$

$\therefore$ $x = 1$

and $y = 0$

Hence the required coordinates of P are (7, 2) or (1, 0).

LOCUS AND ITS EQUATION

Locus

Definition: *The path described or traced by a moving point, when it moves under a given condition or conditions, is called its locus.*

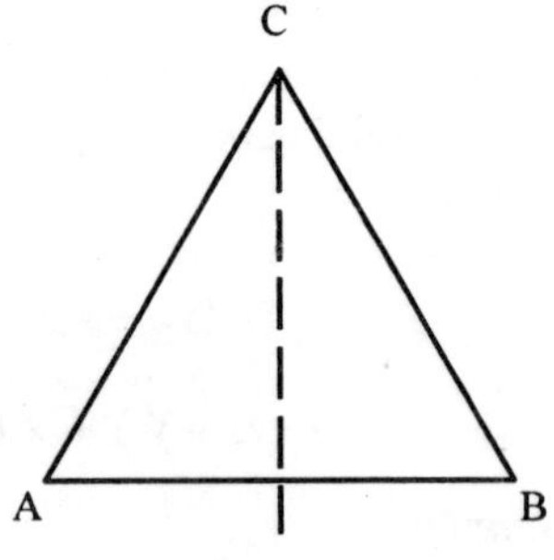

Illustration 1. The locus of a point C which moves so that it remains at equidistant from two given points is the right bisector of the line joining the two given points.

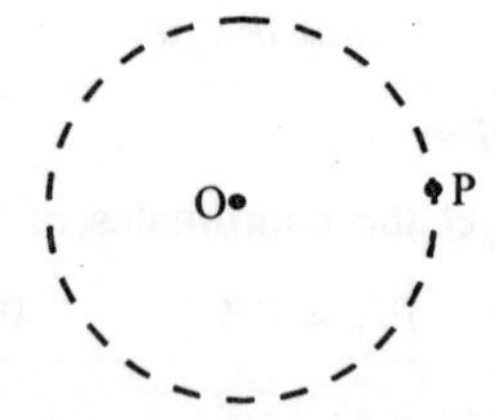

Illustration 2. The locus or path of a point P which moves under the condition that its distance from a fixed point O is always a constant 'k', is the circle whose centre is O and radius is 'k'.

Students are already familiar with the locus in *Plane Geometry* and it has the same definition in Co-ordinate Geometry. In Geometry, we have to find out whether the locus is a straight line or a circle or any other curve.

In Co-ordinate Geometry, if a point moves according to some condition, we can always find an algebraic relation between its co-ordinates. This algebraic relation is called the equation of the locus. In other words, the equation of the locus is defined as an equation is x and y which is satisfied by the co-ordinates of *any* or *every* point on the locus.

As in illustration 2, if a point P(x, y) moves in such a way that its distance from the origin (fixed point O(0, 0) is always k, then its co-ordinates will always satisfy the condition $x^2 + y^2 = k^2$. This represents the equation of the circle with centre O and radius k. Hence $x^2 + y^2 = k^2$ is the equation of the locus of a point.

Example:

The sum of the distances of a point from two fixed points S'(–ae, o) and S(ae, o) is 2a. Find its locus.

Solution:

Let P(x, y) be the moving point. The

$$PS + PS' = 2a$$

$$\therefore \sqrt{(x+ae)^2+y^2} + \sqrt{(x-ae)^2+y^2} = 2a \qquad \text{...(i)}$$

Also $[(x + ae)^2 + y^2] - [(x - ae)^2 + y^2)] = 4aex$ ...(ii)

Dividing (ii) by (i), we get

$$\sqrt{(x+ae)^2+y^2} - \sqrt{(x-ae)^2+y^2} = 2ex. \qquad \text{...(iii)}$$

Adding (i) and (iii), we have

$$2\sqrt{(x+ae)^2+y^2} = 2(a+ex)$$

or $4(x + ae)^2 + y^2 = 4(x + ex)^2$

or $x^2 + y^2 + 2aex + a^2e^2 = a^2 + e^2x^2 + 2aex$

or $(1- e^2)\, x^2 + y^2 = a^2(1 - e^2)$

or $$\frac{x^2}{a^2} + \frac{y^2}{a^2(1-e^2)} = 1.$$

THE STRAIGHT LINES

Slope of a Line

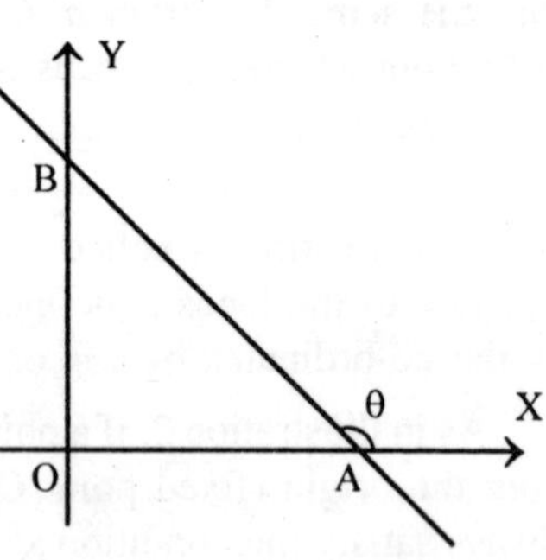

Definition : *The tangent of an angle which a line makes with the positive direction of the x-axis is called the slope of the line.*

It is also called the *gradient* of a line. If θ be the angle that the line aB makes with the positive direction of the x-axis, its slope is tan θ and is generally denoted by m, i.e., tan θ = m. The slope will be positive or negative according as θ is actute or obtuse.

Cor. 1. *When $\theta = 0$, the line is parallel to x-axis and its slope will be tan $0^o = 0$.*

Cor. 2. *When $\theta = 90^o$, the line is perpendicular to x-axis and its slope will be tan $90^o = \infty$. Hence the slope of straight line parallel to x-axis is 0 and parallel to y-axis is ∞.*

EQUATION OF THE STRAIGHT LINES PARALLEL TO THE AXES

To find the equation of a straight line parallel to x-axis.

Proof:

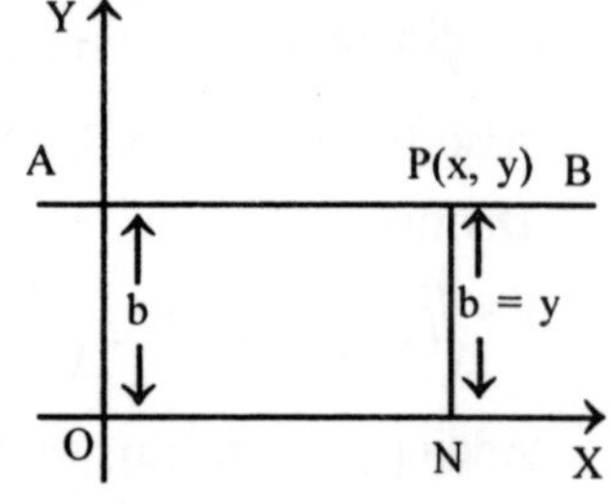

Let AB be a straight line parallel to x-axis and at a distance b from it.

And let P(x, y) be any point on AB and draw PN ⊥ to the x-axis. Then NP = b for all points P on AB. But PN = y.

∴ y = b is the equation of the straight line AB.

If the line AB is below the x-axis, then its equation is y = –b,

When b = 0, the line coincides with x-axis, and get y = 0, the *equation of the x-axis.*

TO FIND THE EQUATION OF A STRAIGHT LINE PARALLEL TO Y-AXIS

Proof:

Let AB be a straight line parallel to y-axis and at a distance a from x-axis. Let P(x, y) be any point on AB and draw PN ⊥ to the y-axis. Then

NP = a for all positions of the point P on the line AB and NP = x. Then we have

x = a is the equation of the line AB.

If the line AB is on the left side of y-axis then its equation is x = – a.

When a = 0, the line concides with y-axis and we get x = 0, the *equation of the y-axis.*

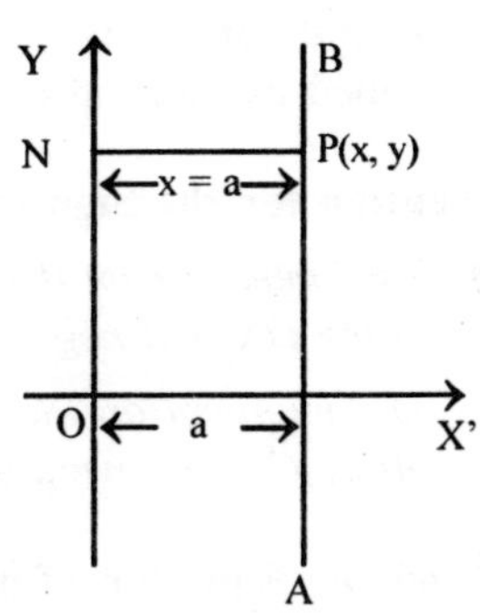

EQUATION OF THE STRAIGHT LINE THROUGH THE ORIGIN

To find the equation of a straight line through the origin making an angle θ with the positive direction of the x-axis.

Proof:

Let AB be any straight line passing through the origin and making an angle θ with the x-axis. Let P(x, y) be any point on AB. Draw PM ⊥ to the x-axis. The OM = x

and MP = y.

From the right-angled ΔOMP, we have

$$\frac{MP}{OM} = \tan\theta$$

or $$\frac{y}{x} = \tan\theta.$$

Writing m for tan θ, the required equation of the line AB is

$$y = mx.$$

Evidently, the line coincides with the x-axis or y-axis according as θ = 0° or 90°.

TANGENT FORM

Intercepts of a Straight Line on the Axes

If a straight line AB meets the x-axis in A and the y-axis in B then :

(i) the distance OA is called the *Intercept* of the line on the x-axis, and

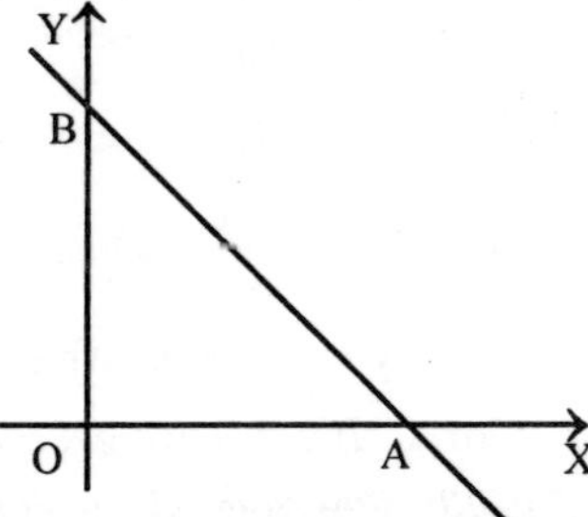

(ii) the distance OB is called the *Intercept* of the line on the y-axis and AB is called the portion of the line intercepted between the axes.

Convention for the Sign of the Intercepts

(i) *The intercept on the x-axis is considered positive, if it is measured along OX and negative if it is measured along OX'.*

(ii) *The intercept on the y-axis is considered positive, if it is measured along OY and negative if it is measured along OY'.*

To Find the Equation of a Straight Line Which Makes an Angle θ With the yx-axis and Which has Intercexpt c on the Y-axis

Proof:

Let AB be the given straight line meeting OY at C so that OC = c and ∠OAB = θ.

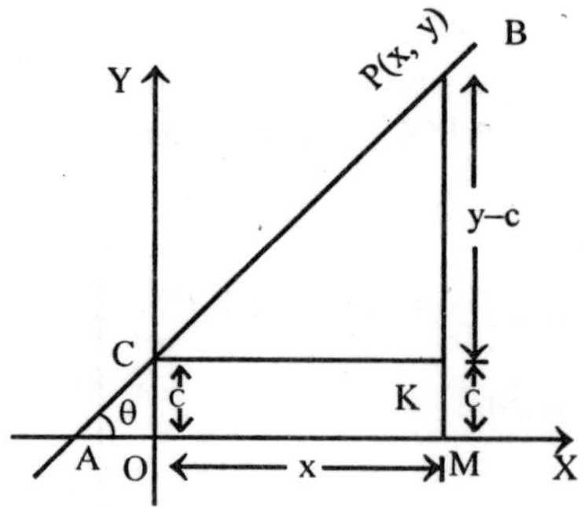

Let P(x, y) be any point on the line AB. Draw PM ⊥ OX and CK ⊥ MP. Then we have

$$\angle KCP = \angle OAB = \theta.$$

Now $\quad CK = OM = x$

and $\quad KP = MP - MK = MP - OC = y - c.$

Now from the right-angled triangle CKP, we have

$$\tan\theta = \frac{KP}{CK}$$

or $\quad m = \dfrac{y-c}{x}$, where $m = \tan\theta$

⇒ $\quad y = mx + c$, which is the required equation.

Cor 1. *If c = 0, we get the equation of the straight line passing through the origin and making an angle θ with x-axis i.e.,* $y = mx$.

Cor 2. *If $m = 0$, the line is parallel to x-axis and the equation of such a line takes the form $y = c$.*

INTERCEPT FORM

To find the equation of a straight line which cuts off given intercepts a and b from the X-axis and Y-axis respectively.

Proof:

Let AB be the given line meeting the axes X and Y in A and B respectively and cutting off the given intercepts a and b. Thus OA = a and OB = b.

Let P(x, y) be any point on AB. Draw PM ⊥ OA.

Now from the similar triangles ΔPMA and ΔBOA we have

$$\frac{MP}{OB} = \frac{MA}{OA}$$

$$= \frac{OA - OM}{OA}$$

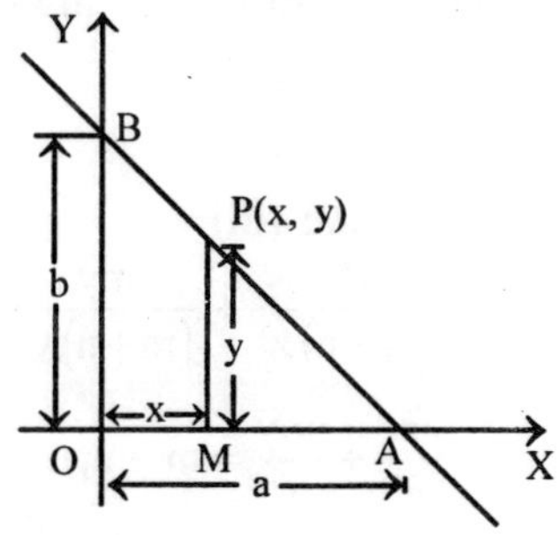

$$\Rightarrow \quad \frac{y}{b} = \frac{a - x}{a} = 1 - \frac{x}{a}$$

$$\text{or} \quad \frac{x}{a} + \frac{y}{b} = 1, \qquad \text{...(i)}$$

which is the required equation of the straight line in intercept form

Notes:

1. *If the intercepts on the axes are given, then the equation of the straight line can at once be written by remembering it in the following form*

$$\frac{x}{\text{Intercepts on } x - \text{axis}} + \frac{y}{\text{Intercepts on } y - \text{axis}} = 1.$$

2. *If in the equation (i) above either a and b approaches infinity, the line becomes parallel to either x-axis or y-axis respectively.*

Example 1:

A straight line passes through the point (x′, y′). Find its equation if its portion intercepted between the axes is divided at (x′, y′) in the rate m : n.

Solution:

Let the euqtion of the line be

$$\frac{x}{a}+\frac{y}{b}=1. \quad ...(i)$$

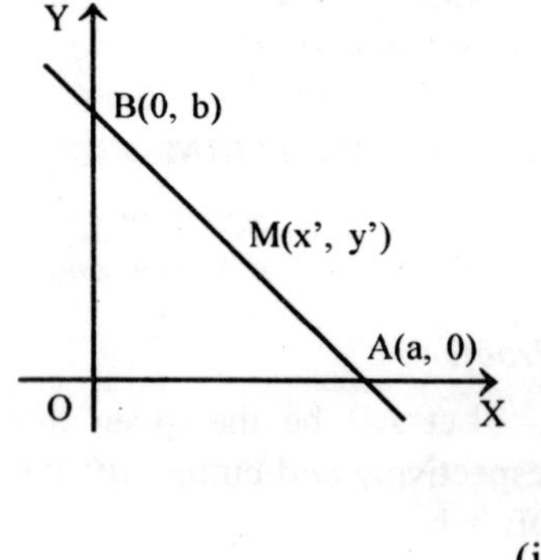

If it meets the axes in A and B, then A is (a, 0) and B is (0, b)

Now (x', y') divides AB in the ratio m : n.

$$\therefore\ x'=\frac{m.o+n.a}{m+n}=\frac{na}{m+n}$$

$$\Rightarrow \qquad a=\frac{(m+n)x}{n} \qquad ...(ii)$$

$$\text{and} \qquad y'=\frac{m.b+n.0}{m+n}=\frac{mb}{m+n}$$

$$\Rightarrow \qquad b=\frac{(m+n)y'}{m} \qquad ...(iii)$$

Now (i), (ii) and (iii)

$$\Rightarrow \qquad \frac{nx}{(m+n)x'}+\frac{my}{(m+n)y'}-1.$$

$$\text{or} \qquad \frac{nx}{x'}+\frac{my}{y'}=m+n,$$

is the required equation.

Example 2:

A straight line is such that the portion of it intercepted between the coordinate axes is bisected at the point (h, k). Prove that its equation is

$$\frac{x}{2h}+\frac{y}{2k}=1.$$

Solution:

Let AB be a straight line and C(h, k) be the middle point of the portion PQ intercepted between the axes.

Let the coordinates of P be (a, 0) and the coordinates of θ be (0, b).

∴ The coordinates of the middle point C of PQ are

$$h=\frac{a+0}{2}$$

and $\quad k = \dfrac{b+0}{2}.$

$\therefore \quad h = \dfrac{a}{2}, k = \dfrac{b}{2}$

$\therefore \quad a = 2h$

and $\quad b = 2k.$

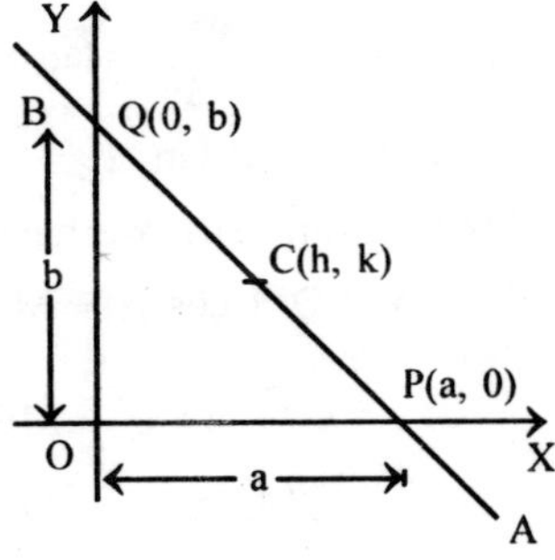

The intercept from of the equation of the straight line AB is

$$\frac{x}{a} + \frac{y}{b} = 1$$

or $$\frac{x}{2h} + \frac{y}{2k} = 1,$$

which is the required equation of the straight line.

NORMAL FORM OR PERPENDICULAR FORM

To find the equation of the straight line in terms of the length of the perpendicular p from the origin upon it and the angle α which this perpendicular makes with the positive direction of x-axis.

Proof:

Let AB be the given straight line and ON be the perpendicular on AB from O, so that ON = p and ∠XON = α. Let P(x, y) be any point on AB. Draw PM ⊥ to the x-axis. Then from the right-angled triangle ΔPMA.We have

∠MAP = 90° – ∠MAP

= ∠AON = α.

(from rt. ∠d ΔONA)

Again, from the rt. ∠d. ΔONA,

We have $\dfrac{ON}{OA} = \cos\alpha.$

∴ ON = OA cos α = (OM + MA) cos α

or p = OM cos α + MA cos α. ...(i)

But we have $\frac{MA}{MP} = \tan\alpha$.

or MA = MP tan α ...(ii)

From (i) and (ii), we have

$$p = OM \cos\alpha + MP. \tan\alpha. \cos\alpha$$

$$= OM \cos\alpha + MP = \frac{\sin\alpha}{\cos\alpha}. \cos\alpha$$

$$= OM \cos\alpha + MP \sin\alpha.$$

$$= x \cos\alpha + y \sin\alpha$$

or $x \cos\alpha + y \sin\alpha - p = 0$,

which is the required equation of a straight line in normal form

Note 1. *The length of the perpendicular p, is always regarded as positive and angle α may have any value between 0° and 360° and is measured in the anti-clockwise direction.*

Note 2.

(i) $\alpha = 0 \Rightarrow x = p$, *which is parallel to x-axis.*

(ii) $\alpha = \frac{\pi}{2} \Rightarrow y = p$, *which is parallel to y-axis.*

(iii) $p = 0 \Rightarrow x \cos\alpha + y \sin\alpha = 0$, *which show that the line passes through the origin.*

(iv) $\alpha = 0$ and $p = 0 \Rightarrow x = 0$, *which is y-axis.*

(v) $\alpha = \frac{\pi}{2}$ and $p = 0 \Rightarrow y = 0$, which is x-axis.

Example 1:

Find the equation of the straight line such that the length of the perpendicular from the origin is of length 4 and the inclination of this perpendicular to the x-axis is 135°.

Solution:

Let the equation of the line be

$$x \cos\alpha + y \sin\alpha = p \qquad ...(i)$$

Here $p = 4$ and $\alpha = 135°$

Substituting the above values in (i), we have

$$x \cos 135° + y \sin 135° = 4$$

or $$x\left(-\frac{1}{\sqrt{2}}\right)+y\left(\frac{1}{\sqrt{2}}\right)=4$$

i.e., $$x - y + 4\sqrt{2} = 0$$

which is the required equation.

Example 2:

To find the equation of the straight line in the perpendicular form assuming the intercept form of the line.

Solution:

Let AB be the given straight line ON, the perpendicular on it from O, so that ON = p and ∠NOX = α.

Let the equation of the line AB be of the form

$$\frac{x}{a}+\frac{y}{b}=1 \qquad \text{...(i)}$$

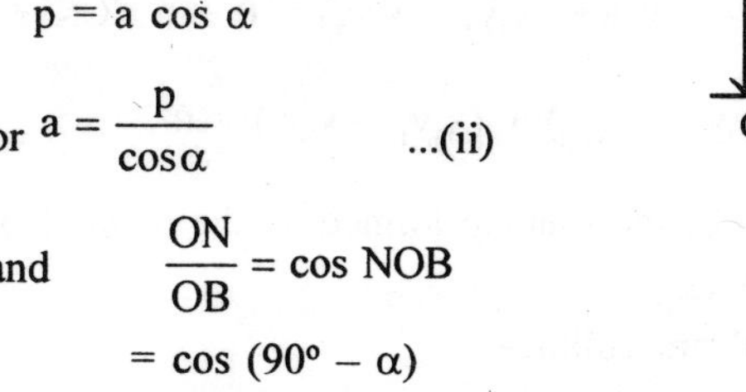

Then $\frac{ON}{OA} = \cos\alpha$

or ON = OA cos α

$p = a\cos\alpha$

or $a = \dfrac{p}{\cos\alpha}$ ...(ii)

and $\dfrac{ON}{OB} = \cos NOB$

$= \cos(90° - \alpha)$

or ON = OB cos (90° – α) = b sin α

or $p = b\sin\alpha$

$$\Rightarrow \quad b = \frac{p}{\cos\alpha}. \qquad \text{...(iii)}$$

Thus, from (i), (ii) and (iii) we have

$$\frac{x}{\frac{p}{\cos\alpha}}+\frac{y}{\frac{p}{\sin\alpha}}=1$$

or $x\cos\alpha + y\sin\alpha = p$.

GENERAL EQUATION OF A STRAIGHT LINE

To Prove that very equation of the first degree in x and y represents a straight line

Proof:

Let the general equation of the first degree in x and y be

$$Ax + By + C = 0. \qquad ...(i)$$

Let $P(x_1, y_1)$, $Q(x_2, y_2)$ and $R(x_3, y_3)$ be any three points on the locus represented by the equation (i).

Since the co-ordinates of these point satisfy equ.(i)

$$\therefore \quad Ax_1 + By_1 + C = 0 \qquad ...(ii)$$

$$Ax_2 + By_2 + C = 0 \qquad ...(iii)$$

$$\text{and} \quad Ax_3 + By_3 + C = 0. \qquad ...(iv)$$

Solving (iii) and (iv) for A, B, C by cross multiplication, we have

$$\frac{A}{y_2 - y_3} = \frac{B}{x_3 - x_2} = \frac{C}{x_2 y_3 - x_3 y_2} = k\,(\text{say}) \qquad ...(v)$$

Substituting, the values of A, B, C from (v) in (ii), we have

$$k\{x_1(y_2 - y_3)\, y_1(x_3 - x_2) + (x_2y_3 - x_3y_2)\} = 0$$

$$\text{or}\,(x_1y_2 - x_2y_1) + (x_2y_3 - x_3y_2) + (x_3y_1 - x_1y_3) = 0 \qquad (CK \neq 0)$$

$$\text{or}\ \frac{1}{2}\,[(x_1y_2 - x_2y_1) + (x_2y_3 - x_3y_2) + (x_3y_1 - x_1y_3) = 0.$$

This shows that the area of the triangle formed by the points $P(x_1, y_1)$ (x_2, y_2) and $R(x_3, y_3)$ is zero.

$\therefore$ The point P, Q and R are collinear.

But these are any three points on the locus of the equation (i). Hence the equation (i) always represents a straight line.

REDUCTION OF THE GENERAL EQUATION OF A STRAIGHT LINE TO THE STANDARD FORMS

(a) Rduction to the Tangent Form

To reduce the equation $Ax + \beta y + C = 0$ to the tangent form of a equation of a line.

Proof:

The given equation is $Ax + By + C = 0$

or $\quad By = -Ax - C$

or $\quad y = -\frac{A}{B}x - \frac{C}{B}$

which is of the form

$$y = mx + c,$$

where $\quad m = -\frac{A}{B}$ and $c = -\frac{C}{B}$.

Note. *Slope of the line $Ax + By + C = 0$ is* $-\frac{A}{B}$

i.e., $\quad -\frac{\text{Co}-\text{eff. of x}}{\text{Co}-\text{eff. of y}}$

and intercept on y-axis is $-\frac{C}{B}$

i.e., $\quad -\frac{\text{Constant term}}{\text{Co}-\text{eff. of y}}$.

(b) Reduction to the Intercept Form

To reduce the equation $Ax + By + C = 0$ to the form $\frac{x}{a} + \frac{y}{b} = 1$.

Proof:

The given euqation is

$$Ax + By + C = 0, \quad C \neq 0$$

or $\quad Ax + By = -C$.

Dividing both sides by $-C$, we have

$$\frac{A}{-C}x + \frac{B}{-C}y = 1$$

or $\quad \frac{x}{\frac{-C}{A}} + \frac{y}{\frac{-C}{B}} = 1$

which is of the form $\frac{x}{a} + \frac{y}{b} = 1$,

where $a = -\frac{C}{A}$ and $b = -\frac{C}{B}$.

(c) Reduction to the Normal Form

To reduce the equation $Ax + By + C = 0$ to the normal form

$x \cos \alpha + y \sin \alpha = p$.

Proof:

The given equation is

$$Ax + By + C = 0. \quad ...(i)$$

Multiply (i) by K(K≠0), we have

$$KAx + KBy + KC = 0. \quad ...(ii)$$

If (ii) be the normal form of (i), then

$$(KA)^2 + (KB)^2 = 1$$

i.e., $$K^2 = \frac{1}{A^2 + B^2}$$

or $$K = \frac{\pm 1}{\sqrt{A^2 + B^2}}$$

Substituting this value of K in (ii), we have

$$\frac{\pm A}{\sqrt{A^2 + B^2}} x + \frac{\pm B}{\sqrt{A^2 + B^2}} y + \frac{\pm C}{\sqrt{A^2 + B^2}} = 0$$

or $$\frac{\pm A}{\sqrt{A^2 + B^2}} x + \frac{\pm B}{\sqrt{A^2 + B^2}} y = -\frac{\pm C}{\sqrt{A^2 + B^2}} \quad ...(iii)$$

The sing of $\pm\sqrt{A^2 + B^2}$ is to be chosen in such a manner so that the expression on the R.H.S. of the (iii) is +ve.

Case I. If C is +ve, the R.H.S. of (iii) will be positive ; if we take -ve sign with the radicokal, the required form is

$$\frac{A}{\sqrt{A^2 + B^2}} x - \frac{B}{\sqrt{A^2 + B^2}} y = \frac{C}{\sqrt{A^2 + B^2}}$$

where $\cos\alpha = -\frac{A}{\sqrt{A^2 + B^2}}$, $\sin\alpha = -\frac{B}{\sqrt{A^2 + B^2}}$

and $$p = \frac{C}{\sqrt{A^2 + B^2}}$$

Case II. If C is –ve, then R.H.S. of (iii) will be positive ; if we take positive sign with the radical, the required form is

$$\frac{A}{\sqrt{A^2 + B^2}} x = + \frac{B}{\sqrt{A^2 + B^2}} y = -\frac{C}{\sqrt{A^2 + B^2}},$$

where $\cos\alpha = \frac{A}{\sqrt{A^2 + B^2}}$, $\sin\alpha = \frac{B}{\sqrt{A^2 + B^2}}$ and $p = \frac{C}{\sqrt{A^2 + B^2}}$,

STANDARD FORMS

One Point Form

To find the equation of a straight line passing through a given point (x_1, y_1) and making a given angle θ with the x-axis.

Proof:

Let m be the slope of the given line AB. Then

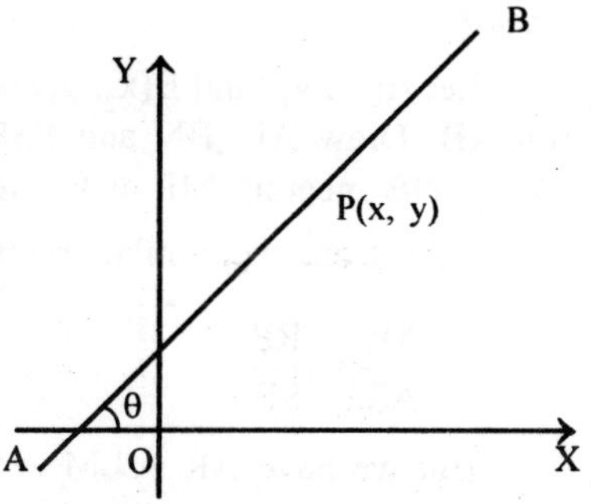

$\therefore m = \tan\theta$

The equation of any line whose slope is m is given by

$y = mx + c$ ...(i)

$\therefore$ It passes through the point (x_1, y_1) then we have

$\therefore y_1 = mx_1 + c$ ...(ii)

subtracting (ii) from (i), we have

$y - y_1 + m(x - x_1)$

which is the required equation of a straight line passing through a given point

TWO POINT FORM

To find the equation of the straight line passing through two points (x_1, y_1) and (x_2, y_2).

Proof:

First Method

The equation of any line through (x_1, y_1) is

$y - y_1 = m(x - x_1)$ ...(i)

where m is the slope of the given line

As it passes through the point (x_2, y_2)

$\therefore y_2 - y_1 = m(x_2, x_1)$

$\Rightarrow m = \frac{y_2 - y_1}{x_2 - x_1}$.

From (i) and (ii), we have

$$y-y_1 = \frac{y_2-y_1}{x_2-x_1}(x-x_1)$$

which is the required equation.

The above equation can also be written in the following form:

$$\frac{y-y_1}{y_2-y_1} = \frac{x-x_1}{x_2-x_1}.$$

Second Method (Direct Method)

Proof:

Let $A(x_1, y_1)$ and $B(x_2, y_2)$ be two given points. Let $P(x, y)$ be any point on AB. Draw AL, BN and PM perpendiculars on OX and from A draw AS ⊥ NB meeting MP at R, we have

Now from the similar triangles ΔARP and ΔASB, we have

$$\frac{AR}{AS} = \frac{RP}{SB} \qquad \text{...(i)}$$

But we have $AR = LM = x - x_1$,

$AS = LN = x_2 - x_1$,

$RP = MP - MR = y - y_1$,

and $SB = NB - NS = y_2 - y_1$.

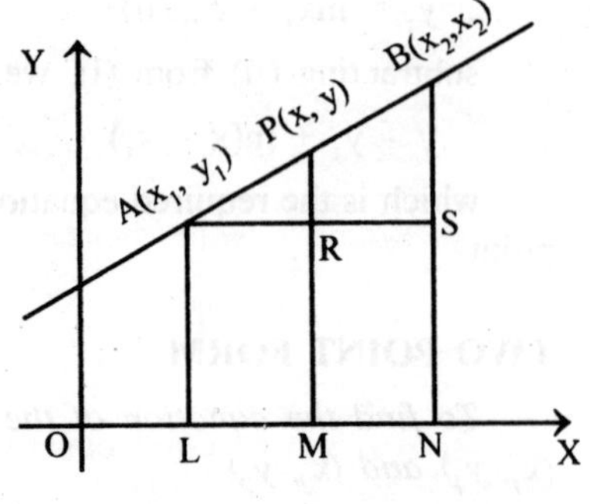

∴ (i) becomes

$$\frac{x-x_1}{x_2-x_1} = \frac{y-y_1}{y_2-y_1}$$

or $y - y_1 = \dfrac{y_2-y_1}{x_2-x_1}(x-x_1)$

which is the required equation.

Note. *Slope of a line joining the two points (x_1, y_1) and (x_2, y_2) is*

$$\frac{y_2-y_1}{x_2-x_1} = \frac{\text{Diff. of y's}}{\text{Diff. of x's}}$$ *taken in the same order.*

SYMMETRY FORM

To find the equation of a straight line passing through a given point (x_1, y_1) and making an angle θ with the x-axis in the form

$$\frac{x-x_1}{\cos\theta} = \frac{y-y_1}{\sin\theta} = r,$$

where r is the distance of any point on the line from the point (x_1, y_1).

Proof:

Let AB be a line passing through the given point $C(x_1, y_1)$ making an angle θ with the x-axis. Let it meet the x-axis at L. Let P(x, y) be any point on AB such that CP = r.

Draw CN and PM perpendicular on the x-axis and $CK \perp$ on MP. Now $\angle KCP = \angle MLC = \theta$.

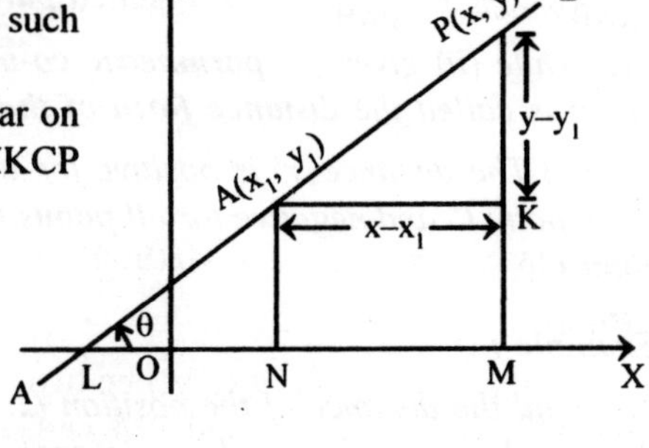

From the right-angled triangle KCP, we have

$$\cos\theta = \frac{CK}{CP}$$

and $$\sin\partial = \frac{KP}{CP}. \qquad \ldots(i)$$

Now $CK = NM = OM - ON = x - x_1$

$KP = MP - MK = MP - NC = y - y_1$

and $CP = r.$

$\therefore$ from (i) we have

$$\cos\theta = \frac{x = x_1}{r} \text{ and } \sin\theta = \frac{y - y_1}{r}$$

or $$\frac{x - x_1}{\cos\theta} = r \text{ and } \frac{y - y_1}{\sin\theta} = r$$

$\therefore$ $$\frac{x - x_1}{\cos\theta} = \frac{y - y_1}{\sin\theta} = r.$$

which is the required equation.

Cor. *To express the co-ordinates of any point on a line passing through a given point and making a given angle with the x-axis in terms of one variable.*

From $\dfrac{x - x_1}{\cos\theta} = \dfrac{y - y_1}{\sin\theta} = r$, we have

$$\left.\begin{aligned} &\frac{x - x_1}{\cos\theta} = r \text{ or } x = x_1 + r\cos\theta \\ \text{and } &\frac{y - y_1}{\sin\theta} = r \text{ or } y = y_1 + r\sin\theta. \end{aligned}\right\} \qquad \ldots(ii)$$

Hence the coordinates of any point on the line are

$(x_1 + r\cos\theta,\ y_1 + r\sin\theta)$.

Note. (i) *The variable r in terms of which the coordinate of any point on the straight line are expressed is called a* ***parameter*** *and the equations* $\frac{x-x_1}{\cos\theta} = r$ and $\frac{y-y_1}{\sin\theta} = r$ *are called* ***parametric equations*** *of the straight line, while (ii) gives the* ***parametric co-ordinates*** *of any point on the line, which is called the* ***distance form of the line.***

(ii) *The distance 'r' is positive for all points lying on one side of the given point C and negative for all points lying on the other side of the given point C.*

Example :

Find the distance of the position (2, 3) from the line $2x - 3y + 9 = 0$ measured along a line making an angle of 45° with the x-axis.

Solution:

The equation of the line through (2, 3) and making an angle of 45° with the x-axis is

$$\frac{x-2}{\cos 45^\circ} = \frac{y-3}{\sin 45^\circ} = r$$

Hence the coordinates of any point on this line at a distance r from (2, 3) is

$(2 + r\cos 45^\circ,\ 3 + r\sin 45^\circ)$

i.e., $\left(2 + \frac{r}{\sqrt{2}},\ 3 + \frac{r}{\sqrt{2}}\right)$

If this point lies on the line $2x - 3y + 9 = 0$, we have

$$2\left(2+\frac{r}{\sqrt{2}}\right) - 3\left(3+\frac{r}{\sqrt{2}}\right) + 9 = 0$$

or $\frac{r}{\sqrt{2}} = 4$ or $r = 4\sqrt{2}$

which is the required distance.

DISTANCE OF A POINT FROM A LINE

To find the perpendicular distance of the point (x_1, y_1) from the line $ax + by + c = 0$.

Proof:

Let PQ be the given line whose equation is given as

$$ax + by + c = 0. \quad ...(i)$$

Let $R(x_1, y_1)$ be any point. Draw RT $\perp$ PQ and join RP and RQ also let RT = p. Then

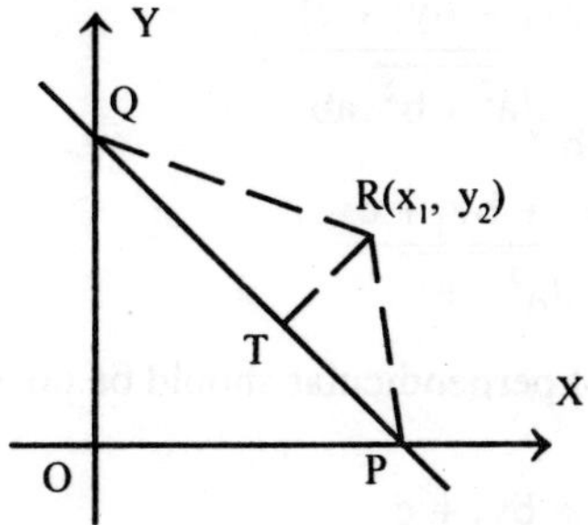

The coordinates of P and Q are $\left(-\frac{c}{a}, 0\right)$ and $\left(0, -\frac{c}{b}\right)$ respectively.

Then by distance formula we have

$$PQ = \sqrt{\left(0+\frac{c}{b}\right)^2 + \left(\frac{-c}{b} - 0\right)^2}$$

$$= \sqrt{\frac{c^2}{a^2} + \frac{c^2}{b^2}} = \frac{c}{ab}\sqrt{a^2 + b^2}$$

Now area of ΔPQR is

$$= \frac{1}{2}\left[x_1\left(-\frac{c}{b} - 0\right) + 0(0 - y_1) - \frac{c}{a}\left(y_1 + \frac{c}{b}\right)\right]$$

$$= -\frac{1}{2}\left(\frac{x_1 c}{b} + \frac{y_1 c}{a} + \frac{c^2}{ab}\right)$$

$$= \frac{-c}{2ab}(ax_1 + by_1 + c) \quad ...(i)$$

Also area of the ΔPQR = $\frac{1}{2}$ base × height

$$= \frac{1}{2}\text{ PQ.p.} \quad ...(ii)$$

From (i) and (ii), we have

$$\frac{1}{2}\,PQ.p = -\frac{c}{2ab}\left(ax_1 + by_1 + c\right)$$

$$p = -\frac{c}{PQ}\frac{\left(ax_1 + by_1 + c\right)}{ab}$$

$$= -\frac{c\left(ax_1 + by_1 + c\right)}{\frac{c}{ab}\sqrt{a^2+b^2}.ab}$$

$$= -\frac{ax_1 + by_1 + c}{\sqrt{a^2+b^2}}$$

Since, the length of perpendicular should be positive, the required length p is given by

$$d = \frac{ax_1 + by_1 + c}{\sqrt{a^2+b^2}}.$$

Example :

Show that if the three points

$\left(\frac{a^3}{a-1}, \frac{a^2-3}{a-1}\right), \left(\frac{b^3}{b-1}, \frac{b^2-3}{b-1}\right), \left(\frac{c^3}{c-1}, \frac{c^2-3}{c-1}\right)$ *are collinear then*

abc – (bc + ca + ab) + 3 (a + b + c) = 0.

Solution:

Let the given three points be collinear on the line

$lx + my + 1 = 0.$ ...(i)

Since the given three points lie on the (1),

$$\therefore \left.\begin{aligned} &l\left(\frac{a^3}{a-1}\right) + m\left(\frac{a^2-3}{a-1}\right) + 1 = 0 \\ &l\left(\frac{b^3}{b-1}\right) + m\left(\frac{b^2-3}{b-1}\right) + 1 = 0 \\ &l\left(\frac{c^3}{c-1}\right) + m\left(\frac{c^2-3}{c-1}\right) + 1 = 0 \end{aligned}\right\} \quad \text{(ii)}$$

The three equation given by (2) show that a, b, c are the roots of the equation

$$l\left(\frac{x^3}{x-1}\right)+m\left(\frac{x^2-3}{x-1}\right)+1=0$$

or $$lx^3 + mx^2 + x - (3m + 1) = 0$$

Hence from the theory of equations, we have

$$a + b + c = -\frac{m}{l},$$

$$ab + bc + ca = \frac{1}{l},$$

$$abc = \frac{3m+1}{l}$$

$$\therefore\ abc - (bc + ca + ab) + 3(a + b + c) = \frac{3m+1}{l} - \frac{1}{l} + 3\left(-\frac{m}{l}\right)$$

$$= 0.$$

MISCELLANEOUS EXAMPLES

Example 1:

Find the distance between the following pairs of points (a cos α, a sin α) and (a cos β, a sin β).

Solution:

The points are (a cos α, a sin α) and (a cos β, a sin β). Hence the required distance

$$= \sqrt{\{(a\cos\alpha - a\cos\beta)^2 + (a\sin\alpha - a\sin\beta)^2\}}$$

$$= \sqrt{\left\{\left(a.2\sin\frac{\alpha+\beta}{2}\sin\frac{\beta-\alpha}{2}\right)^2 + \left(a.2\cos\frac{\alpha+\beta}{2}\sin\frac{\alpha-\beta}{2}\right)^2\right\}}$$

$$= \sqrt{\left\{4a^2\sin^2\frac{\alpha+\beta}{2}\sin^2\frac{\alpha-\beta}{2} + 4a^2\cos^2\frac{\alpha+\beta}{2}\sin^2\frac{\alpha-\beta}{2}\right\}}$$

$$= \sqrt{\left\{4a^2 . \sin^2\frac{\alpha+\beta}{2} . \left(\sin^2\frac{\alpha+\beta}{2} + \cos^2\frac{\alpha+\beta}{2}\right)\right\}}$$

$$= \sqrt{\left(4a^2 . \sin^2\frac{\alpha+\beta}{2}\right)} = 2a.\sin\frac{\alpha-\beta}{2}$$ **Ans.**

Example 2:

The line joining the points (–6, 8) and (8, –6) is divided into four equal parts; find the co-ordinates of the points of section.

Solution:

Let the points (–6, 8) and (8, –6) be A and B respectively. If P is the mid-point of AB, the co-ordinates of P are given by

$$\left(\frac{-6+8}{2}, \frac{8-6}{2}\right) \text{ or } (1, 1)$$

$$\left[\because \text{ Mid-point of } (x_1, y_1), (x_2, y_2) \text{ is } \frac{x_1 + x_2}{2}, \frac{y_1 + y_2}{2}\right]$$

Again if Q be the mid-point of AP, the co-ordinates of Q will be

$$\left(\frac{-6+1}{2}, \frac{8+1}{2}\right) \text{ or } \left(-\frac{5}{2}, \frac{9}{2}\right)$$

If R be the middle point of PB, the co-ordinates of R will be

$$\left(\frac{1+8}{2}, \frac{1-6}{2}\right) \text{ or } \left(\frac{9}{2}, \frac{5}{2}\right)$$

Hence the required points are $\left(-\frac{5}{2}, \frac{9}{2}\right)$; (1, 1) or $\left(\frac{9}{2}, -\frac{5}{2}\right)$ **Ans.**

Example 3:

Prove that the lines joining the middle points of opposite sides of a quadrilateral and the line joining the middle points of its diagonals meet in a point and bisect one another.

Solution:

Let ABCD be the quad. having the co-ordinates of the vertices as (x_1, y_1), (x_2, y_2), (x_3, y_3) and (x_4, y_4) respectively.

Let P, Q, R and S be the middle points of AB, BC, CD and DA respectively, and L, M be the middle points of the diagonals BD and AC respectively. Then the co-ordinates of P are

$$\left(\frac{x_1 + x_2}{2}, \frac{y_1 + y_2}{2}\right),$$

Similarly the point Q is $\left(\frac{x_2 + x_3}{2}, \frac{y_2 + y_3}{2}\right)$

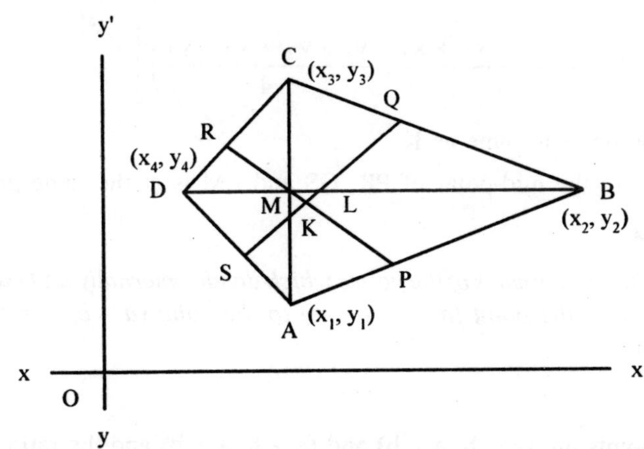

R is $\left(\dfrac{x_3 + x_4}{2}, \dfrac{y_3 + y_4}{2}\right)$, S is $\left(\dfrac{x_4 + x_1}{2}, \dfrac{y_4 + y_1}{2}\right)$,

L is $\left(\dfrac{x_2 + x_4}{2}, \dfrac{y_2 + y_4}{2}\right)$ and M is $\left(\dfrac{x_1 + x_3}{2}, \dfrac{y_1 + y_3}{2}\right)$

Again let the mid-point of PR be K. Then the co-ordinates of K are

$$\left[\frac{1}{2}\left(\frac{x_1 + x_2}{2} + \frac{x_1 + x_4}{2}\right), \frac{1}{2}\left(\frac{y_1 + y_2}{2} + \frac{y_3 + y_4}{2}\right)\right]$$

or $\left(\dfrac{x_1 + x_2 + x_3 + x_4}{4}, \dfrac{y_1 + y_2 + y_3 + y_4}{4}\right)$

Similarly the mid-point of SQ is

$$\frac{1}{2}\left(\frac{x_1 + x_4}{2} + \frac{x_2 + x_3}{2}\right), \frac{1}{2}\left(\frac{y_1 + y_4}{2} + \frac{y_2 + y_3}{2}\right)$$

or $\left(\dfrac{x_1 + x_2 + x_3 + x_4}{4}, \dfrac{y_1 + y_2 + y_3 + y_4}{4}\right)$

Therefore the mid-point of QS is also K.

Mid point of LM is

$$\left[\frac{1}{2}\left(\frac{x_2 + x_4}{2} + \frac{x_3 + x_1}{2}\right), \frac{1}{2}\left(\frac{y_2 + y_4}{2} + \frac{y_3 + y_1}{2}\right)\right]$$

or $$\left[\left(\frac{x_1 + x_2 + x_3 + x_4}{4}, \frac{y_1 + y_2 + y_3 + y_4}{4}\right)\right]$$

So this point is same as K.

Therefore the mid-point of PR, QS and LM is at the same point K.

Example 4:

Find the co-ordinates of the points which divide, internally and externally, the line joining the point (a + b, a – b) to the point (a – b, a = b) in the ratio a : b.

Solution:

The points are (a + b, a – b) and (a – b, a + b) and the ratio is a : b. Hence for internal division the required co-ordinates are

$$\left[\frac{a(a-b) + b\,(a+b)}{a+b}, \frac{a(a+b) + b\,(a-b)}{a+b}\right]$$

or $$\left(\frac{a^2 + b^2}{a+b}, \frac{a^2 + 2ab - b^2}{a+b}\right)$$ **Ans.**

For external division, the ratio will be a : –b so the co-ordinates of the required point are

$$\left[\frac{a(a-b) + (-b)\,(a+b)}{a-b}, \frac{a(a+b) + (-b)\,(a-b)}{a-b}\right]$$

or $$\left(\frac{a^2 - 2ab - b^2}{a-b}, \frac{a^2 + b^2}{a-b}\right)$$ **Ans.**

Example 5:

Find the co-ordinates of the point which divides the line joining the points (1, 3) and (2, 7) in the ratio 3 : 4.

Solution:

Points are (1, 3) and (2, 7). The ratio is 3 : 4. Hence the required co-ordinates are

$$\left[\frac{(3 \times 2) + (4 \times 1)}{3+4}, \frac{(3 \times 7) + (4 \times 3)}{3+4}\right] \text{ or } \left(\tfrac{10}{7}, \tfrac{33}{7}\right).$$ **Ans.**

Example 6:

Find the co-ordinates of point which divides the same line in the ratio 3 : –4.

Solution:

The points are (1, 3) and (2, 7) and the ratio is (3 : –4). Hence the required co-ordinates are

$$\left[\frac{(3 \times 2) + (-4 \times 1)}{3 - 4}, \frac{(3 \times 7) + (-4 \times 3)}{3 - 4}\right] \text{ or } (-2, -9). \qquad \textbf{Ans.}$$

Example 7:

If G be the centroid of a triangle ABC and O be any other point, prove that $3\,(GA^2 + GB^2 + GC^2) = BC^2 + CA^2 + AB^2$.

and $OA^2 + OB^2 + OC^2 = GA^2 + GB^2 + GC^2 + 3GO^2$.

Solution:

Let us suppose that the axes of co-ordinates pass through the centroid G of the triangle ABC. Let the co-ordinates of A, B and C be (x_1, y_1), (x_2, y_2) and (x_3, y_3) respectively.

Then the centroid is given by $\left(\frac{x_1 + x_2 + x_3}{3}, \frac{y_1 + y_2 + y_3}{3}\right)$ which is supposed to be at the origin.

Hence $x_1 + x_2 + x_3 = 0$ ... (1) and $y_1 + y_2 + y_3 = 0$ ...(2)

We have to prove that $3\,(GA^2 + GB^2 + GC^2) = BC^2 + CA^2 + AB^2$.

Now $GA^2 = (x_1 - 0)^2 + (y_1 - 0)^2 = x_1^2 + y_1^2$.

Similarly, $GB^2 = x_2^2 + y_2^2$ and $GC^2 = x_3^2 + y_3^2$.

Hence $\text{L.H.S.} = 3\,(GA^2 + GB^2 + GC^2)$

$$= 3\,(x_1^2 + x_2^2 + x_3^2 + y_1^2 + y_2^2 + y_3^2) \qquad ...(3)$$

And $BC^2 = (x_2 - x_3)^2 + (y_2 - y_3)^2$

$$= x_2^2 + x_3^2 - 2x_2x_3 + y_2^2 + y_3^2 + 2y_2y_3$$

Similarly, $CA^2 = x_3^2 + x_1^2 - 2x_3x_1 + y_3^2 + y_1^2 - 2y_3y_1$

and $AB^2 = x_1^2 + x_2^2 - 2x_1x_2 + y_1^2 + y_2^2 - 2y_1y_2$

Therefore $\text{R.H.S.} = BC^2 + CA^2 + AB^2$

$$= 2\,(x_1^2 + x_2^2 + x_3^2 + y_1^2 + y_2^2 - y_3^2)$$

$$-2\,(x_1x_2 + x_2x_3 + x_3x_1 + y_1y_2 + y_2y_3 + y_3y_1) \qquad ...(4)$$

By equation No. (1), we have $x_1 + x_2 + x_3 = 0$.

On squaring, we get $x_1^2 + x_2^2 + x_3^2 + 2x_1x_2 + 2x_2x_3 + 2x_3x_1 = 0$

or $\quad x_1^2 + x_2^2 + x_3^2 = -2(x_1x_2 + x_2x_3 + x_3x_1)$ ...(5)

Similarly be relation (2), we get

$y_1^2 + y_2^2 + y_3^2 = -2(y_1y_2 + y_2y_3 + y_3y_1)$ ...(6)

Adding (5) and (6), we get, $(x_1^2 + x_2^2 + x_3^2 + y_1^2 + y_2^2 + y_3^2)$

$= -2(x_1x_2 + x_2x_3 + x_3x_1 + y_1y_2 + y_2y_3 + y_3y_1)$

Substituting the value of R.H.S. in equation (4), we get

$BC^2 + CA^2 + AB^2 = 2(x_1^2 + x_2^2 + x_3^2 + y_1^2 + y_2^2 + y_3^2)$

$+ (x_1^2 + x_2^2 + x_3^2 + y_1^2 + y_2^2 + y_3^2)$

$= 3(x_1^2 + x_2^2 + x_3^2 + y_1^2 + y_2^2 + y_3^2)$

[by relation (3)]

= L.H.S. **Hence proved.**

Again let O be any point (x, y), then

$OA^2 = (x - x_1)^2 + (y - y_1)^2 = x^2 + x_1^2 - 2xx_1 + y^2 + y_1^2 = 2yy_1$

Similarly, $\quad OB^2 = x^2 + x_2^2 - 2xx_2 + y^2 + y_2^2 - 2yy_2$

and $\quad OC^2 = x^2 + x_3^2 - 2xx_3 + y^2 + y_2^2 - 2yy_3$

Therefore L.H.S. $= OA^2 + OB^2 + OC^2$.

$= 3x^2 + x_1^2 + x_3^2 + x_2^2 + 3y^2 + y_1^2 + y_2^2 + y_3^2$

$-2x(x_1 + x_2 + x_3) - 2y(y_1 + y_2 + y_3)$

$= 3(x^2 + y^2) + (x_1^2 + y_1^2) + (x_2^2 + y_2^2) + (x_3^2 + y_3^2)$ ...(7)

$\because x_1 + x_2 + x3 = y_1 + y_2 + y_3 = 0$ [by equations (1), (2)]

Now $\quad GA^2 = (x_1 - 0)^2 + (y_1 - 0)^2 = (x_1^2 + y_1^2)$

Similarly $\quad GB^2 = (x_2^2 + y_2^2)$, $GC^2 = (x_3^2 + y_3^2)$

and $\quad GO^2 = (x^2 + y^2)$, $\quad \therefore \quad 3GO^2 = 3(x^2 + y^2)$.

R.H.S. $= GA^2 + GB^2 + GC^2 + 3GO^2$

$= (x_1^2 + y_1^2) + (x_2^2 + y_2^2) + (x_3^2 + y_3^2) + 3(x^2 + y^2)$

= L.H.S. [by equation (7)]. **Hence proved.**

Example 8:

The line joining the points (1, –2) and (–3, 4) is trisected; find the co-ordinates of the points of trisection.

Solution:

Let the *points* (1, –2) and (–3, 4) be A and B respectively. If P and Q be the points of trisection of AB, the ratio AP : PB will be equal to 1 : 2 and the ratio AQ : QB will be equal to 2 : 1. Hence the co-ordinates of P are

$$\left[\frac{(1\times-3)+(2\times1)}{1+2}, \frac{(1\times4)+(2\times-2)}{1+2}\right] \text{ or } \left(\frac{1}{3}, 0\right) \qquad \textbf{Ans.}$$

Co-ordinates of Q are

$$\left[\frac{(2\times-3)+(1\times1)}{2+1}, \frac{(2\times4)+(1\times-2)}{2+1}\right] \text{ or } \left(\frac{5}{3}, 2\right) \qquad \textbf{Ans.}$$

Example 9:

Find the co-ordinates of the point which divides, internally and externally, the line joining (–1, 2) to (4, –5) in the ratio 2 : 3.

Solution:

The points are (1, 2) and (4, –5). The ratio is 2 : 3.

(i) For internal division, the required co-ordinates are

$$\left[\frac{(2\times4)+(3\times-1)}{2+3}, \frac{(2\times-5)+(3\times2)}{2-3}\right] \text{ or } \left(1, \frac{4}{5}\right) \qquad \textbf{Ans.}$$

(ii) For external division, the ratio will be 2 : – 3, so the required co-ordinates are

$$\left[\frac{(2\times4)+(-3\times-1)}{2-3}, \frac{(2\times-5)+(-3\times2)}{2-3}\right] \text{ or } (11, 16) \qquad \textbf{Ans.}$$

Example 10:

Find the co-ordinates of the point which divides, internally and externally, the line joining (–3 –4), to (–8, 7) in the ratio 7 : 5.

Solution:

The points are (–3, –4) to (–8, 7) and the ratio is 7 : 5.

(i) For internal division, the co-ordinates of the required points are

$$\left[\frac{(7\times-8)+(5\times-3)}{7+5}, \frac{(7\times7)+(5\times-4)}{7+5}\right] \text{ or } \left(-\frac{71}{12}, \frac{29}{12}\right) \qquad \textbf{Ans.}$$

(ii) For external division, the ratio will be 7 : – 5; so the co-ordinates of the required points are

$$\left[\frac{(7 \times -8) + (-5 \times -3)}{7 - 5}, \frac{(7 \times 7) + (-5 \times -4)}{7 - 5}\right] \text{ or } \left(\frac{41}{2}, \frac{69}{2}\right) \quad \textbf{Ans.}$$

Example 11:

Prove that the points (2, –2), (8, 4), (5, 7) and (–1, 1) are at the angular points of a rectangle.

Solution:

Let the points (2, –2), (8, 4), (5, 7) and (–1, 1) be A, B, C and D respectively.

Then, Side AB = $\sqrt{[(2 - 8)^2 + (-2 - 4)^2]} = \sqrt{[(36 + 36)} = 6\sqrt{2}$, ...(1)

Side BC = $\sqrt{[(8 - 5)^2 + (4 - 7)^2]} = \sqrt{[(9 + 9)} = 3\sqrt{2}$, ...(2)

Side CD = $\sqrt{[(5 + 1)^2 + (7 - 1)^2]} = \sqrt{[(36 + 36)} = 6\sqrt{2}$, ...(3)

Side DA = $\sqrt{[(-1 - 2)^2 + (1 + 2)^2]} = \sqrt{[(9 + 9)} = 3\sqrt{2}$, ...(4)

Diagonal AC = $\sqrt{[(2 - 5)^2 + (-2 - 7)^2]} = \sqrt{[(9 + 81)} = 3\sqrt{10}$,...(5)

Diagonal BD = $\sqrt{[(8 + 1)^2 + (4 - 1)^2]} = \sqrt{[(81 + 9)} = 3\sqrt{10}$, ...(6)

From equations (1) to (6) we find that AB = CD, BC = AD and AC = BD. Hence ABCD is a quad. in which opposite sides are equal and the diagonals are also equal, hence a rectangle.

Example 12:

Find the distance between the , following pairs of points (2, 3) and (5, 7).

Solution:

The points are (2, 3) and (5, 7). Hence the required distance

$$= \sqrt{\{(2 - 5)^2 + (3 - 7)^2\}} + \sqrt{(9 + 16)} = 5$$

Example 13:

Find the distance between the following pairs of points (4, –7) and (–1, 5).

Solution:

The points are (4, –7) and (–1, 5). Hence the required distance $= \sqrt{\{(4) - (-1)\}^2 + \{(-7)^2 - (5)\}^2]}$ **Ans.**

Example 14:

Find the distance between the following pairs of points (–3, –2) and (–6, 7), the axes being inclined at 60°.

Solution:

The points are (–3, –2) and (–6, 7) and $\omega = \cos 60°$. Hence the required distance $= \sqrt{[\{(-3) - (-6)\}^2 + \{(-2) - (7)\}^2 + 2\{(-3) - (-6)\}\{(-2) - (7)\} \cos 60°]}$

$$= \sqrt{\left[\{-3 + 6\}^2 + \{-2 - 7\}^2 + 2(-3 + 6\}(-2 - 7).\frac{1}{2}\right]}$$

$$= \sqrt{\left\{9 + 81 + 2.3(-9)\frac{1}{2}\right\}} = \sqrt{[(63)} = 3\sqrt{7}.$$ **Ans.**

Example 15:

Prove that the points (2a, 4a), (2a, 6a) and $(2a + \sqrt{3}a, 5a)$ are the vertices of an equilateral triangle whose side is 2a.

Solution:

Let the points (2a, 4a); (2a, 6a) and $[2a + \sqrt{3}a, 5a]$ be A, B and C respectively. Then

$AB = \sqrt{[(2a - 2a)^2 + (4a - 6a)^2]} = 2a,$ ...(1)

$AC = \sqrt{[(2a + \sqrt{3}a - 2a)^2 + (5a - 4a)^2]} = \sqrt{(3a^2 + a^2)} = 2a,$ ...(2)

$BC = \sqrt{[(2a + \sqrt{3}a - 2a)^2 + (5a - 6a)^2]} = \sqrt{(3a^2 + a^2)} = 2a.$ ...(3)

By (1), (2) and (3), we get AB = BC = Ca = 2a. **Proved.**

Example 16:

Prove that the points (–2, –1), (1, 0), (4, 3) and (1, 2) are at the vertices of a parallelogram.

Solution:

Prove the points (–2, –1), (1, 0), (4, 3) and (1, 2) be A, B, C and D respectively; then

$AB = \sqrt{[(-2 - 1)^2 + (-1 - 0)^2]} = \sqrt{(9 + 1)} = \sqrt{10}$ ...(1)

$BC = \sqrt{[(1 - 4)^2 + (0 - 3)^2]} = \sqrt{(9 + 9)} = 3\sqrt{2}$ ...(2)

$CD = \sqrt{[(4 + 1)^2 + (3 - 2)^2]} = \sqrt{(9 + 1)} = \sqrt{10}$ ...(3)

$DA = \sqrt{[(-2 - 1)^2 + (-1 - 2)^2]} = \sqrt{(9 + 9)} = 3\sqrt{2}$ ...(4)

From (1), (2), (3) and (4), we get AB = CD and BC = DA. Hence, the opposite sides being equal, the quadrilateral is a parallelogram.

Example 17:

Find the distance between the following pairs of points $(am_1^2, 2am_1)$ *and* $(am_2^2, 2am_2)$.

Solution:

The points are $(am_1^2, 2am_1)$ and $(am_2^2, 2am_2)$. Hence the required distance
$= \sqrt{\{(am_1^2 - am_2^2)^2 + (2am_1 - 2am_2)^2\}}$

$= \sqrt{\{a^2 (m_1 - m_2)^2 (m_1 + m_2)^2 + 4a^2 (m_1 - m_2)^2\}}$

$= \sqrt{[a^2 (m_1 - m_2)^2 + \{(m_1 - m_2)^2 + 4\}]}$

$= a (m_1 - m_2) \sqrt{\{(m_1 - m_2)^2 + 4\}}$. **Ans.**

Example 18:

Lay down in a figure the positions of the points (1, –3) and (–2, 1) and prove that the distance between them is 5.

Solution:

The points are (1, –3) and (–2, 1). Hence the distance between them
$= \sqrt{[\{(1) - (-2)^2 + \{(-3) - (1)\}^2]} = \sqrt{\{(3)^2 + (-4)^2\}}$
$= \sqrt{(9 + 16)^2} = 5.$ **Proved.**

Example 19:

Find the value of x_1 *if the distance between the points* $(x_1, 2)$ *and (3, 4) be 8.*

Solution:

The points are $(x_1, 2)$ and (3, 4). Hence the distance between them
$= \sqrt{\{(x_1 - 3)^2 + (2 - 4)^2\}} = 8$ (by hypothesis).
on squaring both sides and simplifying, we get
$x_1^2 + 9 - 6x_1 + 4 = 64$ or $x_1^2 - 6x_1 - 51 = 0$

$$\text{Hence } x_1 = \frac{\pm\sqrt{[36 - 4.1(-51)}}{2} = \frac{6 + 2\sqrt{60}}{2} = 3 \pm 2\sqrt{15}.$$

Example 20:

A line is of length 10 and one end is at the point (2, –3); if the abscissa of the other end be 10, prove that its ordinate must be 3 or –9.

Solution:

Let the ordinate of the second point be y. So the first point is (2, –3) and second is (10, y). The distance between them (i.e. the length of the line) $= \sqrt{[(2 = 10)^2 + (3 + y)^2]} = 10$ (by hypothesis)

or $64 + y^2 = 9 + 6y = 100$ (on squaring and simplifying)

or $y^2 + 6y = 27 - 0$ or $(y - 3)(y + 9) = 0$. ∴ $y = 3$ or -9. **Proved.**

Example 21:

Prove that the point $\left(-\frac{1}{14}, \frac{39}{14}\right)$ *is the centre of the circle circumscribing the triangle whose angular points are (1, 1), (2, 3), and (–2, 2).*

Solution:

Let the points (1, 1), (2, 3) and (–2, 2) be A, B and C respectively.

Let us suppose P be the circumcentre whose co-ordinates are (x, y)

Hence ...(1)

Now $OA^2 = (x - 1)^2 + (y - 1)^2 = x^2 + y^2 - 2x - 2y + 2,$ ...(2)

$OB^2 = (x - 2)^2 + (y - 3)^2 = x^2 + y^2 - 4x - 6y + 13,$ ...(3)

$OC^2 = (x - 2)^2 + (y - 2)^2 = x^2 + y^2 + 4x - 4y + 8,$ ...(4)

By (1), $OA^2 = OB^2$

So $x^2 + y^2 - 2x - 2y + 2 = x^2 + y^2 + 4x - 6y + 13$

or $2x + 4y = 11$ ...(5)

Again by (1), $OB^2 = OC^2$.

So $x^2 + y^2 - 4x - 6y + 13 = x^2 + y^2 + 4x - 4y + 8$

or $-8x - 2y = -5$ or $8x + 2y = 5$ ...(6)

Solving equations (5) and (6), we get $= -\frac{1}{14}, y = \frac{39}{14}$

Hence the co-ordinates of the circumcentre P are $\left(-\frac{1}{14}, \frac{39}{14}\right)$. **Proved**

Example 22:

Find the distance between the following pairs of points (a, 0) and (0, b).

Solution:

The points are (a, 0) and (0, b). Hence the required distance $= \sqrt{\{(a = 0)^2 + (0 - b^2)\}} = \sqrt{(a^2 + b^2)}$. **Ans.**

Example 23:

Find the equations to the straight lines which pass through the origin and are inclined at 75° to the straight line $x + y + \sqrt{3}(y - x) = a$.

Solution:

(a) $$y - 0 = \frac{(\sqrt{3}-1)/\sqrt{3}+1 + \tan 75°}{1 - (\sqrt{3}-1)/(\sqrt{3}+1).\tan 75°}(x - 0)$$

$$\left(\because \text{m of the given equation is } \frac{\sqrt{3}-1}{\sqrt{3}+1}\right)$$

$$\Rightarrow \quad y = \frac{\frac{\sqrt{3}-1}{\sqrt{3}+1} + \frac{\sqrt{3}+1}{\sqrt{3}-1}}{1 - \frac{\sqrt{3}-1}{\sqrt{3}+1}.\frac{\sqrt{3}+1}{\sqrt{3}-1}}.x \quad \left(\because \tan 75° = \frac{\sqrt{3}+1}{\sqrt{3}-1}\right)$$

$\Rightarrow \quad x = 0.$ **Ans.**

(b) $$y - 0 = \frac{\frac{\sqrt{3}-1}{\sqrt{3}+1} - \tan 75°}{1 + \frac{\sqrt{3}-1}{\sqrt{3}+1}.\tan 75°}(x - 0)$$

$$\Rightarrow \quad y = \frac{\frac{\sqrt{3}-1}{\sqrt{3}+1} - \frac{\sqrt{3}+1}{\sqrt{3}-1}}{1 + \frac{\sqrt{3}-1}{\sqrt{3}+1}.\frac{\sqrt{3}+1}{\sqrt{3}-1}}.x \text{ or } y + \sqrt{3}.\ x = 0.$$ **Ans.**

Example 24(a):

Find the angles between the pairs of straight lines

$(m^2 - mn)\ y = (mn + n^2)\ x + n^3$

and $(mn + m^2)\ y = (mn - n^2)\ x + m^3$.

Solution:

The equations of the lines are $(m^2 - mn)$

$y = (mn + n^2)\ x + n^2$

and $(mn + m^2)\ y = (mn - n^2)\ x + m^3$.

Therefore their respective slopes are $\frac{mn + n^2}{m^2 - mn}$ and $\frac{mn - n^2}{mn + n^2}$

$$\text{Hence } \tan\theta = \frac{\dfrac{mn+n^2}{m^2-mn} - \dfrac{mn-n^2}{mn+n^2}}{1+\dfrac{mn+n^2}{m^2-mn}\times\dfrac{mn-n^2}{mn+n^2}}$$

$$= \frac{\left(mn+n^2\right)\left(mn+m^2\right)-\left(mn-n^2\right)\left(m^2-mn\right)}{\left(m^2-mn\right)\left(m^2+mn\right)+\left(mn+n^2\right)\left(mn-n^2\right)}$$

$$= \frac{nm\,(m+n^2) - n.m\,(m-n)^2}{m^2\,(m-n)\,(m+n) + n^2\,(m-n)\,(m-n)}$$

$$= \frac{nm\;4mn}{(m^2-n^2)\,(m^2+n^2)} = \frac{4m^2n^2}{m^4-n^4}$$

$$\theta = \tan^{-1}\left[\frac{4m^2n^2}{m^2-n^2}\right]$$

Example 24(b):

Find the equation to the straight line drawn at right angles to the straight line x/a – y/b = 1 through the point where it meets the axis of x.

Solution:

The given line is $x/a - y/b = 1$. It is clear that it cuts x–axis at (a, 0) and its slope is b/a.

Slope of the line perpendicular to it is –a/b (By $m_1m_2 = -1$.)

Hence equation of the line passing through (a, 0) and having slope – a/b, will be

$$y - 0 = -a/b\,(x - a) \quad \text{or} \quad ax + by = a^2.$$

Example 24(c):

Find the equation to the straight line which bisects, and is perpendicular to the straight line joining the points (a, b) and (a', b').

Solution:

The slop of the line joining (a, b) and (a', b') is $= \dfrac{b'-b}{a'-a}$

Hence slope of the line perpendicular to it will be $= -\dfrac{a'-a}{b'-b}$

The mid point of the line (a, b) and (a', b') is $\left(\dfrac{a+a'}{2}, \dfrac{b+b'}{2}\right)$.

Hence the required equation is $\left(y - \dfrac{b - b'}{2}\right) = \dfrac{a' - a}{b' - b}\left(x - \dfrac{a - a'}{2}\right).$

Simplifying, we get, $2(a - a')x + 2(b - b')y = a^2 - a'^2 + b^2 - b'^2$ **Ans.**

Example 24(d):

Find the angles between the pairs of straight lines $y = 3x + 7$ and $3y - x = 8$.

Solution:

The lines are $y = 3x + 7$ and $3y - x = 8$.

Therefore $m_1 = 3$ and $m_2 = \dfrac{1}{3}$.

Hence $\tan\theta = \dfrac{3 - \frac{1}{3}}{1 + 3.\frac{1}{3}} = \dfrac{\frac{8}{3}}{2} = \dfrac{4}{3}$ $\therefore \theta = \tan^{-1}\left(\frac{4}{3}\right)$ **Ans.**

Example 25:

Find the angles between the pairs of straight lines $y = (2 - \sqrt{3})x + 5$ and $y = (2 + \sqrt{3})x - 7$.

Solution:

The lines are $y = (2 - \sqrt{3})x + 5$ and $y = (2 + \sqrt{3})x - 7$.

Here $m_1 = 2\sqrt{3}$ and $m_2 = 2 + \sqrt{3}$.

Therefore $\tan\theta = \dfrac{(2 - \sqrt{3}) - (2 + \sqrt{3})}{1 + (2 - \sqrt{3}).(2 + \sqrt{3})} = -\dfrac{2\sqrt{3}}{2} = -\sqrt{3}$

or $\tan\theta = \sqrt{3}$ (omitting –ve sign).

$\theta = 60°$. **Ans.**

Example 26:

Find the tangent of the angle between the lines whose intercepts on the axes are respectively a, –b and b –a.

Solution:

The intercepts on the axes by the first line are a and –b. Therefore the equation of the line is $x/a + y/{-b} = 1$ or $ay - bx + ab = 0$.

Similarly the intercepts by the second line are (b, – a).

Hence its equation is $x/b + y/{-a}$ or $by - ax + ab = 0$.

The values of m_1 and m_2 are respectively b/a and a/b.

So $$\tan\theta = \frac{b/a - a/b}{1 + b/a\ a/b} = -\frac{b^2 - a^2}{2ab}$$

or $$\theta = \tan^{-1}\left(\frac{a^2 - b^2}{2ab}\right).$$ (omitting –ve sign).

Example 27(a):

Prove that the points (2, –1), (0, 2), (2, 3) and (4, 0) are the co-ordinates of the angular points of a parallelogram and find the angle between its diagonals.

Solution:

Let the given points (2, –1), (0, 2), (2, 3) and (4, 0) be A, B, C and D respectively.

co-ordinates of the middle points of the diagonal AC are

$$\left(\frac{2+2}{2}\right), \left(\frac{3-1}{2}\right), \text{ or } (2, 1).$$

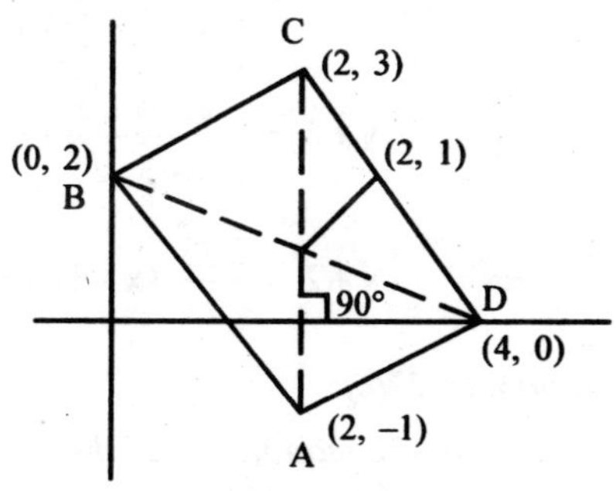

Co-ordinates of the middle points of the diagonal BD are

$$\left(\frac{0+4}{2}\right), \left(\frac{2+0}{2}\right) \text{ or } (2, 1)$$

Therefore the two diagonals bisect each other. Hence the quad, must be a parallelogram.

Now as the slope of the line joining (x_1, y_1) is $= \frac{y_2 - y_1}{x_2 - x_1}$, we have the slope of diagonal AC $= \frac{3+1}{2-2} = \infty$, and slope of the diagonal BD $= \frac{2-0}{0-4} = -\frac{1}{2}$.

The first diagonal is parallel to y-axis as the slope;

i.e. $\tan\theta = \infty$ or $\theta = 90°$.

∴ The angle between AC and BD = $\tan^{-1}$ (2). **Ans.**

[complementary to the acute angle that the line BD makes with axis of x, i.e. $\tan^{-1}\left(\frac{1}{2}\right)$].

Example 27(b):

Show that the equations to the straight lines passing through the point (3, –2) and inclined at 60° to the line $\sqrt{3}x + y = 1$ are $y + 2 = 0$ and $y - \sqrt{3}x + 2 + 3\sqrt{3} = 0$.

Solution:

The equations of the straight lines passing through (3, –2) and inclined at an angle of 60° to the line

$\sqrt{3}x + y = 1$

are (a) $y+2 = \dfrac{-\sqrt{3} + \tan 60°}{1 + \sqrt{3}\tan 60°}(x-3)$ (∵ m of the given line is $-\sqrt{3}$)

$$\Rightarrow \quad y+2 = \frac{1 - \sqrt{3} + \sqrt{3}}{1 - (-\sqrt{3})\sqrt{3}}(x-3) \quad \text{or } y + 2 = 0. \qquad \textbf{Ans.}$$

(b) $y+2 = \dfrac{-\sqrt{3} - \tan 60°}{1 + (-\sqrt{3})\tan 60°}(x-3)$

$$\Rightarrow \quad y+2 = \frac{-\sqrt{3} + \sqrt{3}}{1 - \sqrt{3}.\sqrt{3}}(x-3)$$

$$\Rightarrow \quad y+2 = \frac{2\sqrt{3}}{2}(x-3) \quad \text{or } y - \sqrt{3}x + 2 + 3\sqrt{3} = 0. \qquad \textbf{Ans.}$$

Example 27(c):

Find the equation to the straight line passing through the point (–6, 10) and perpendicular to the straight line $7x + 8y = 5$.

Solution:

The given line is, $7x + 8y = 5$. ...(1)

any line perpendicular to this line will be $8x - 7y = \lambda$. ...(2)

(where λ is any constant)

As the required line passes through (–6, 10), the co-ordinates will satisfy (2), hence $8(-6) - 7(10) = \lambda$ or $\lambda = -118$.

Substituting in (2), we get $8x - 7y + 118 = 0$. **Ans.**

Example 28(a):

Find the equation to the straight line passing through the point (2, –3) and perpendicular to the straight line joining the points (5, 7) and (–6, 3).

Solution:

Equation of any line passing through the given point

(2, –3) is (y + 3) = m(x – 2). ...(1)

'm' of the straight line joining the point (5, 7) and (–6, 3) is

$$= \frac{3-7}{-6-5} = \frac{4}{11} \quad \left(\therefore\ m = \frac{y_2 - y_1}{x_2 - x_1}\right) \qquad ...(2)$$

As the lines are perpendicular $m \times \left(\frac{4}{11}\right) = -1$, ($\because$ $m_1\ m_1 = -1$ of p)

or $m = -\dfrac{4}{11}$.

Substitution in (1) and simplifying, we get

$(y + 3) = -\dfrac{4}{11}(x - 2)$ or $4y + 11x = 10$ **Ans.**

Example 28(b):

Through the point (3, 4) are drawn two straight lines each inclined at 45° to the straight line x – y = 2. Find their equations and find also the area included by the three lines.

Solution:

The equations of the straight lines passing through (3, 4) and making an angle 45° to the straight line

x – y = 2. ...(1)

will be

(a) $y - 4 = \dfrac{1 + \tan 45°}{1 - 1.\tan 45°}(x - 3)$

$\therefore$ 'm' of the given line is (1)

or $y - 4 = \dfrac{1+1}{1-1}(x - 3)$

or x – 3 = 0 ...(2) **Ans.**

(b) $y - 4 = \dfrac{1 - \tan 45°}{1 + 1.\tan 45°}(x - 3)$

or $y - 4 = \dfrac{1-1}{1+1}(x - 3)$

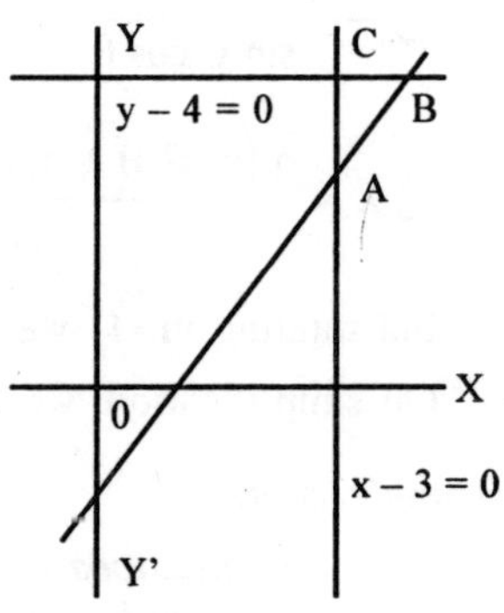

$\Rightarrow \quad y - 4 = 0 \quad ...(3)$ **Ans.**

Let the lines given by (1), (2) and (3) meet at A, B and C.

Solving (1) and (2), we get the co-ordinates of A and (3, 1).

Solving (1) and (3), we get the co-ordinates of B as (3, 4).

Solving (2) and (3), we get the co-ordinates of C as (3, 4).

As x = 3 and y = 4 are parallel to axis of y and axis of x, respectively, ABC is a triangle right-angled at C.

Again CA = 3 and CB = 3.

So area of $\Delta ABC = \frac{1}{2} \times 3 \times 3 = \frac{9}{2}$ units.

Example 28(c):

Prove that the equation to the straight line which passes through the point ($a \cos^3 \theta$, $a \sin^3 \theta$) and is perpendicular to the straight line $x \sec \theta + y \operatorname{cosec} \theta = a$ is $x \cos \theta - y \sin \theta = a \cos 2\theta$.

Solution:

Any line perpendicular to the line $x \sec \theta + y \operatorname{cosec} \theta = a$ may be given by $x \operatorname{cosec} \theta - y \sec \theta = \lambda$ (where λ is a constant) ...(1)

As the required line passes through ($a \cos^3 \theta$, $a \sin^3 \theta$), so it will satisfy (1).

Hence $a \cos^3 \theta \,.\, \operatorname{cosec} \theta - a \sin^3 \theta \,.\, \sec \theta = \lambda$

$$\Rightarrow \frac{a \cos^3 \theta}{\sin \theta} - \frac{a \sin^3 \theta}{\cos \theta} = \lambda$$

$$\Rightarrow \frac{a \cos^4 \theta - a \sin^4 \theta}{\sin \theta \cos \theta} = \lambda$$

$$\Rightarrow \lambda = \frac{a\left(\cos^2 \theta + \sin^2 \theta\right)\left(\cos^2 \theta - \sin^2 \theta\right)}{\sin \theta \cos \theta} = \frac{a \cos 2\theta}{\sin \theta \cos \theta}$$

Substituting in (1) we get $x \cos \theta - y \sec \theta = a \cos 2\theta / \sin \theta \cos \theta$.

On simplification, we get, $x \cos \theta - y \sin \theta = a \cos 2\theta$.

Example 28(d):

Find the equations to the straight lines passing through (x', y') and respectively perpendicular to the straight lines $xx' + yy' = a^2$, $xx'/a^2 + yy'/b^2 = 1$, and $x'y + xy' = a^2$.

Solution:

Any line passing through the point (x', y') may be given by

$y - y' = m(x - x')$. ...(1)

(a) As (1) is perpendicular to it, slop of (1) is $(+y'/x') = m$.

Putting in (1), we get $y - y' = \frac{y'}{x'}(x - x')$

$\Rightarrow \quad x'y - yx' = 0$. **Ans.**

(b) The given line is $\frac{xx'}{a^2} + \frac{yy'}{b^2} = 1$, hence its slope is $\left(-\frac{x'/a^2}{y'/b^2}\right)$

$\therefore$ The slope of the line perpendicular to it is $\left(+\frac{y'/b^2}{x'/a^2}\right) = \frac{a^2y'}{b^2x'}$.

Putting it in (1), we get $y - y' = \frac{a^2y'}{b^2x'}(x - x')$.

(c) The given line is $x'y + xy' = a^2$. $\therefore$ slope $= (-y'/x')$.

So the slope of the line perpendicular to it is $(+x'/y')$.

Putting it in (1), we get $y - y' = \frac{x'}{y'}(x - x')$

$\Rightarrow xx' - yy' = x'^2 - y'^2$. **Ans.**

Example 29:

Find the equations to the straight lines which divide, internally and externally, the line joining (–3, 7) to (5, –4) in the ratio of 4 : 7 and which are perpendicular to this line.

Solution:

Let the given points (–3, 7) and (5, –4) be A and B respectively. The co-ordinates of the point which divides AB in the ration 4 : 7 internally will be

$$\left[\frac{4 \times 5 + 7 \times (-3)}{4+7}, \frac{4 \times (-4) + 7 \times 7}{4 + 8}\right] \text{ or } \left(-\frac{1}{11}, 3\right)$$

Slope of line AB $= \frac{-4 - 7}{5-(-3)} = -\frac{11}{8}$.

Hence slope of the line perpendicular to it will be $= \frac{8}{11}$.

Equation of the line passing through $\left(-\frac{1}{8}, 3\right)$ and perpendicular to AB is $y - 3 = \frac{8}{11}\left(x + \frac{1}{11}\right)$

$121y - 88x = 371.$ **Ans.**

Again let the point P (x_1, y_1) divide AB in the ratio 4 : 7 externally. Then

$$x_1 = \frac{(4)(5) - (7)(-3)}{4 - 7} = -\frac{41}{3};\ y_1 = \frac{(4 \times -4) - (7)(7)}{4 - 7} = +\frac{65}{3}$$

Hence the line passing through $P = \left(-\frac{41}{3}, \frac{65}{3}\right)$ and perpendicular to AB is

$$y - \frac{65}{3} = \frac{8}{11}\left(x + \frac{41}{3}\right) = \frac{8}{33}(3x + 41) \qquad \because m = \frac{8}{11} \text{ by (1)}$$

$33y - 24x = 1043.$ **Ans.**

Example 30(a):

Show that the product of the perpendiculars drawn from the two points $(\pm \sqrt{a^2 - b^2}, 0)$ upon the straight line $\frac{x}{a} \cos \theta + \frac{y}{a} \sin \theta = 1$ is b^2.

Solution:

The points are $[\pm \sqrt{a^2 - b^2}), 0]$ and the straight line is

$$\frac{x}{a} \cos \theta + \frac{y}{a} \sin \theta = 1$$

$\Rightarrow \qquad x\, b \cos \theta + y\, a \sin \theta - ab = 0.$

If p be the length of the perpendicular from $[\sqrt{(a^2 - b^2)}, 0]$ and p' be the length of the perpendicular from $[-\sqrt{(a^2 - b^2)}, 0]$ upon the line given by (1), we have

$$p = \frac{(b \cos \theta) \sqrt{(a^2 - b^2)} - (a \sin \theta) \times 0 - ab}{\sqrt{(b^2 \cos^2 \theta + a^2 \sin^2 \theta)}}$$

$$= \frac{b\left[(\cos \theta) \sqrt{(a^2 - b^2)} - a\right]}{\sqrt{(b^2 \cos^2 \theta + a^2 \sin^2 \theta)}}$$

and $$p' = \frac{(b\cos\theta)\left[-\sqrt{(a^2-b^2)}\right] + (a\sin\theta)\times 0 - ab}{\sqrt{(b^2\cos^2\theta + a^2\sin^2\theta)}}$$

$$= -\frac{-b\left[(\cos\theta)\sqrt{(a^2-b^2)}.-a\right]}{\sqrt{(b^2\cos^2\theta + a^2\sin^2\theta)}}$$

$$\therefore \quad pp' = \frac{b\left[(\cos\theta)\sqrt{(a^2-b^2)}-a\right]}{\sqrt{(b^2\cos^2\theta + a^2\sin^2\theta)}} \times \frac{-b(\cos\theta)\left[\sqrt{(a^2-b^2)}\right]+a}{\sqrt{(b^2\cos^2\theta + a^2\sin^2\theta)}}$$

$$= \frac{b^2}{(b^2\cos^2\theta + a^2\sin^2\theta)} \times \left[\cos^2\theta\,(a^2-b^2) - a^2\right]$$

$$= \frac{b^2}{(b^2\cos^2\theta + a^2\sin^2\theta)}\left[a^2\cos^2\theta - b^2\cos^2\theta - a^2\right]$$

$$= \frac{b^2}{(b^2\cos^2\theta + a^2\sin^2\theta)}\left[-b^2\cos^2\theta - a^2(1-\cos^2\theta)\right]$$

$$= +\frac{b^2}{(b^2\cos^2\theta + a^2\sin^2\theta)}\left[b^2\cos^2\theta + a^2\sin^2\theta\right]$$

$= b^2$. **Proved.**

Example 30(b):

Find the length of the perpendicular from the origin upon the straight line joining the two points whose co-ordinates are (a cos α, a sin α) and (a cos β, a sin β).

Solution:

The point is (0, 0). The line is the one joining

(a cos α, a sin α) and (a cos β, a sin β).

Hence the equation of line is

$$y - a\sin\beta = \frac{a\sin\alpha - a\sin\beta}{a\cos\alpha - a\cos\beta}(x - a\cos\beta)$$

$$\Rightarrow x\cos\left(\frac{\alpha+\beta}{2}\right) + y\sin\left(\frac{\alpha+\beta}{2}\right) - a\cos\frac{1}{2}(\alpha-\beta) \;\; 0.$$

Therefore length of perpendicular

$$= \frac{0 \times \cos \frac{1}{2}(\alpha + \beta) + 0 \times \sin \frac{1}{2}(\alpha + \beta) - a \cos \frac{1}{2}(\alpha - \beta)}{\sqrt{\left[\cos^2 \left\{\frac{1}{2}(\alpha + \beta)\right\} + \sin^2 \left\{\frac{1}{2}(\alpha + \beta)\right\}\right]}}$$

$$= a \cos \frac{1}{2}(\alpha - \beta).$$

[$\because \cos^2 \theta + \sin^2 \theta = 1$ and omitting the –ve sign]. **Ans.**

Example 31:

Find the length of the perpendicular drawn from the point (4, 5) upon the straight line 3x + 4y = 10.

Solution:

The point is (4, 5) and the line is

$3x + 4y = 10$ or $3x + 4y - 10 = 0$.

Hence the length of perpendicular $= \dfrac{3.4 + 4.5 - 10}{\sqrt{(3^2 + 4^2)}} = \dfrac{22}{5} = 4\dfrac{2}{5}$ unit.

Ans.

Example 32:

Find the length of the perpendicular drawn from the origin upon the straight line x/3 – y/4 = 1.

Solution:

The point origin i.e. (0, 0), the line is

$x/3 - y/4 = 1$ or $4x - 3y - 12 = 0$

Hence the length of perpendicular $= \dfrac{4.0 - 3.0 - 12}{\sqrt{(4^2 + 3^2)}} = -\dfrac{2}{5} = 2\dfrac{2}{5}$ units.

(neglecting the sign.) **Ans.**

Example 33:

Find the length of the perpendicular drawn from the point (–3, –4) upon the straight line 12(x + 6) = 5(y – 2).

Solution:

The point is (–3, –4); the straight line is

$12(x + 6) = 5(y - 2)$ or $12x - 5y + 82 = 0$.

Hence length of perpendicular

$$= \frac{12(-3) - 5(-4) + 82}{\sqrt{(12^2 + 5^2)}} = \frac{66}{13} = 5\frac{1}{13} \text{ units.}$$

Example 34:

Find the length of the perpendicular drawn from the point (b, a) upon the straight line x/a = y/b = 1.

Solution:

The point is (b, a), straight line is x/a – y/b = 1.

$\Rightarrow \quad ay - bx + ab = 0.$

Hence the length of perpendicular

$$= \frac{-b.b + a.a + ab}{\sqrt{(b^2 + a^2)}} = \frac{a^2 + ab - b^2}{\sqrt{(a^2 + b^2)}} \quad \textbf{Ans.}$$

Example 35(a):

If p and p' be the perpendiculars from the origin upon the straight lines whose equations are x sec θ + y cosec θ = a and x cos θ – y sin θ = a cos 2θ, prove that $4p^2 + p'^2 = a^2$.

Solution:

The point is given to be the origin (0, 0) and lines are

$$x \sec\theta + y \operatorname{cosec}\theta = a \quad ...(1)$$

and $\quad x \cos\theta - y \sin\theta = a \cos 2\theta. \quad ...(2)$

If p and p' are the perpendiculars from (0, 0) on (1) and (2) respectively, then

$$p = \frac{0.\sec\theta + 0.\operatorname{cosec}^2\theta - a}{\sqrt{(\sec^2\theta + \operatorname{cosec}^2\theta)}} = -\frac{a}{\sqrt{(\sec^2\theta + \operatorname{cosec}^2\theta)}} \quad ...(3)$$

and $$p' = \frac{0.\cos\theta - 0.\sin\theta - a\cos 2\theta}{\sqrt{(\cos^2\theta + \sin^2\theta)}} = -\frac{a\cos 2\theta}{1} \quad ...(4)$$

Now, $$4p^2 + p'^2 = \frac{4a^2}{\sec^2\theta + \operatorname{cosec}^2\theta} + a^2\cos^2 2\theta$$

(putting the values of p and p')

$$= \frac{4a^2}{\frac{1}{\cos^2\theta} + \frac{1}{\sin^2\theta}} + a^2\cos^2 2\theta = \frac{4a^2\cos^2\theta\sin^2\theta}{\sin^2\theta + \cos^2\theta} + a^2\cos^2 2\theta$$

$$= a^2 \sin^2 2\theta + a^2 \cos^2 2\theta = a^2 (\sin^2 2\theta + \cos^2 2\theta)$$

$$= a^2$$

Ans.

Example 35(b):

Find the distance between the two parallel straight lines y = mx + c and y = mx + d.

Solution:

Let the given lines y = mx + c and y = mx + d be AB and CD respectively.

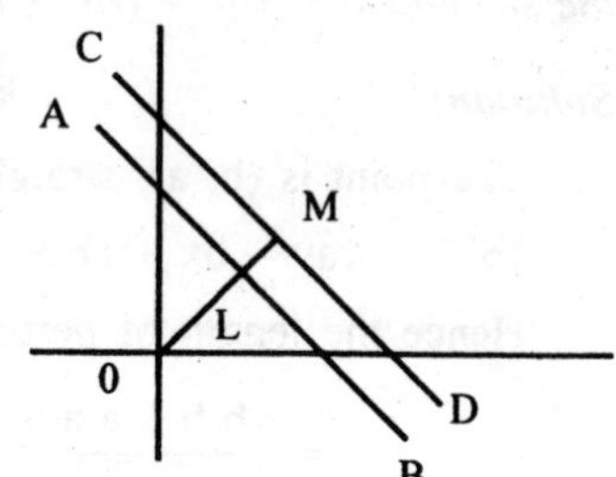

If OL and OM be the perpendiculars from the origin on AB and CD respectively, then, clearly, the distance between the lines is

$$LM = OM - OL. \quad ...(1)$$

Now $$OM = \frac{0.m + 0.1 - d}{\sqrt{(m^2 + 1)}} = \frac{-d}{\sqrt{(m^2 + 1)}}.$$

$$OL = \frac{0.m + 0.1 - c}{\sqrt{(m^2 + 1)}} = \frac{-c}{\sqrt{(m^2 + 1)}}$$

Therefore, $$ML = OM - OL = \frac{-d}{\sqrt{(m^2 + 1)}} - \frac{-c}{\sqrt{(m^2 + 1)}} = \frac{(c - d)}{\sqrt{(m^2 + 1)}}.$$

Example 36:

What are the points on the axis of x whose perpendicular distance from the straight line x/a + y/b = 1 is a?

Solution:

The line is x/a + y/b = 1 or bx + ay – ab = 0. ...(1)

Any point on x-axis can be taken as (h, 0), hence by hypothesis, the perpendicular from (h, 0) on (1) is equal to a

$\Rightarrow$ The perpendicular distance $= \dfrac{h.b + 0.a - ab}{\sqrt{(a^2 + b^2)}} = \pm a$

($\because$ the distance may be +ve or –ve in sign)

$\Rightarrow$ $hb - ab = \pm \sqrt{(a^2 + b^2)}$

$$\Rightarrow \qquad h = \frac{1}{b}\left[ab \pm a\sqrt{(a^2 + b^2)}\right]$$

$$= \frac{a}{b}\left[b \pm \sqrt{(a^2 + b^2)}\right]$$

Hence the points are $\left[\frac{a}{b}\left\{b \pm \sqrt{(a^2 + b^2)}\right\}, 0\right]$. **Ans.**

Example 37:

Show that the perpendiculars let fall from any point of the straight line $2x + 11y = 5$ upon the two straight lines $24x + 7y = 20$ and $4x - 3y = 2$ are equal to each other.

Solution:

Let (h, k) be any point from which the perpendiculars drawn to the given lines $24x + 7y = 20$ ...(1)

and $4x - 3y = 2$

are equal in length.

Th length of perpendicular from (h, k) on (1)

$$= \frac{24h + 7k - 20}{\sqrt{(24^2 + 7^2)}} = \frac{24h + 7k - 20}{25} \qquad ...(3)$$

and the length of perpendicular from (h, k) on (2)

$$= \frac{4h - 3k - 2}{\sqrt{(4^2 + 3^2)}} = \frac{4h - 3k - 2}{5} \qquad ...(4)$$

By hypothesis, (3) and (4) are equal; so

$$= \frac{24h + 7k - 20}{25} = \frac{4h - 3k - 2}{5}$$

$$\Rightarrow \qquad 4h + 22k = 10$$

$$\text{or} \qquad 2h + 11k = 5. \qquad ...(5)$$

Generalising equation (5), we get the locus of the point from which the perpendiculars let fall on (1) and (2) are equal.

Hence the locus is $2x + 11y = 5$ or conversely from any point on $2x + 11y = 5$, the length of perpendiculars on (1) and (2) are equal. **Proved.**

Example 38:

Find the distance between the following pairs of points (b + c, c + a) and (c + a, a + b).

Solution:

The points are (b + c, c + a) ; (c + a, a + b). Hence the required distance

$= \sqrt{[\{(b + c) - (c + a)\}^2] + \{(c + a) - (a + b)\}^2]}$

$= \sqrt{\{(b - a)^2 + (c - b)^2\}} = \sqrt{(b^2 + a^2 - 2ab + c^2 - b^2 - 2cb)}$

$= \sqrt{\{a^2 + 2b^2 + c^2 - 2ab - 2ac\}}$. **Ans.**

Example 39(a):

A, B, C, D.... are n points in a place whose co-ordinates are (x_1, y_1), (x_2, y_2), (x_3, y_3), ... AB is bisected in the point G_1, G_1C is divided at G_2 in the ratio 1 : 2, G_2D is divided at G_3 in the ratio 2 : 3; G_3E at G_4 in the ratio 1 : 4, and so on until all the points are exhausted. Show that the co-ordinates of the final point so obtained are

$$\frac{x_1 + x_2 + x_3 + ... + x_n}{n} \text{ and } \frac{y_1 + y_2 + y_3 + ... + y_n}{n}$$

| *This point is called the* ***Centre of Mean Position*** *of the n given points.* |

Solution:

The co-ordinates of the points A and B are given to be (x_1, y_1) and (x_2, y_2).

The co-ordinates of G_1, the mid-point of AB, will be

$$\left(\frac{x_1 + x_2}{2}, \frac{y_1 + y_2}{2}\right)$$

The co-ordinates of 3rd point C are (x_3, y_3)

Co-ordinates of G_2 (which divides G_1C in the ratio of 1 : 2 will be

$$\left[\frac{2\left(\frac{x_1 + x_2}{2}\right) + x_3}{2 + 1}, \frac{2\left(\frac{y_1 + y_2}{2}\right) + y_3}{2 + 1}\right]$$

or $$\left(\frac{x_1 + x_2 + x_3}{3}, \frac{y_1 + y_2 + y_3}{3}\right)$$

The co-ordinates of the point D are (x_4, y_4).

The co-ordinates of G_3 (which divides G_2D in the ratio 1 : 3) will be

$$\left\{\frac{3\dfrac{x_1+x_2+x_3}{3}+x_4}{3+1}, \frac{3\dfrac{y_1+y_2+y_3}{3}+y_4}{3+1}\right\}$$

or $$\left(\frac{x_1+x_2+x_3+x_4}{4}, \frac{y_1+y_2+y_3+y_4}{4}\right)$$

Proceeding in the same way (n – 1) times, we get the centre of mean position of n given points as

$$\left[\frac{1}{n}(x_1+x_2+x_3+\ldots+x_n), \frac{1}{n}(y_1+y_2+y_3+\ldots+y_n)\right]$$ **Proved.**

Example 39(b):

Prove that a point can be found which is at the same distance from each of the four points

$$\left(am_1, \frac{a}{m_1}\right), \left(am_2, \frac{a}{m_2}\right), \left(am_3, \frac{a}{m_3}\right) \text{ and } \left(\frac{a}{m_1 m_2 m_3}, am_1 m_2 m_3\right)$$

Solution:

Let the given points

$$\left(am_1, \frac{a}{m_1}\right), \left(am_2, \frac{a}{m_2}\right) \text{ and } \left(am_3, \frac{a}{m_3}\right)$$

be A, B and C respectively. Suppose any point O having the co-ordinates (x, y) is such that OA = OB = OC ...(1)

Now, $OA^2 = (x-am_1)^2+\left(y-\frac{a}{m_1}\right)^2$, $OB^2 = (x-am_2)^2+\left(y-\frac{a}{m^2}\right)^2$,

$OC^2 = (x-am_3)^2+\left(y-\frac{a}{m_3}\right)^2$

By (1), we get

$$(x-am_1)^2+\left(y-\frac{a}{m_1}\right)^2 = (x-am_2)^2+\left(y-\frac{a}{m_2}\right)^2 \quad ...(2)$$

and $$(x-am_1)^2+\left(y-\frac{a}{m_1}\right)^2 = (x-am_3)^2+\left(y-\frac{a}{m_3}\right)^2 \quad ...(3)$$

Solving (2) and (3), we get

$$x = \frac{a}{2}\left(m_1 + m_2 + m_3 + \frac{1}{m_1 m_2 m_3}\right)$$

$$y = \frac{a}{2}\left(\frac{1}{m_1} + \frac{1}{m_2} + \frac{1}{m_3} + \frac{1}{m_1 m_2 m_3}\right)$$

Putting these values of x and y in the value of OA^2, we get

$$OA^2 = \left[\left\{\frac{a}{2}\left(m_1 + m_2 + m_3 + \frac{1}{m_1 m_2 m_3}\right) - am_1\right\}^2 + \left\{\frac{a}{2}\left(\frac{1}{m_1} + \frac{1}{m_2} + \frac{1}{m_3} + m_1 m_2 m_3\right) - \frac{a}{m_1}\right\}^2\right]$$

$$= \frac{a^2}{4}\left\{m_1^2 + m_2^2 + m_3^2 + \frac{1}{m_1^2} + \frac{1}{m_2^2} + \frac{1}{m_3^2} + m_1^2 + m_2^2 + m_3^2 + \frac{1}{m_1^2 + m_2^2 + m_3^2}\right\} \quad ...(4)$$

Let D be any point having co-ordinates $= \left(\frac{a}{m_1 m_2 m_3}, am_1 m_2 m_3\right)$

$$\text{Then } OD^2 = \left[\left\{\frac{a}{2}\left(m_1 + m_2 + m_3 + \frac{1}{m_1 m_2 m_3}\right) - \frac{a}{m_1 m_2 m_3}\right\}^2 + \left\{\frac{a}{2}\left(\frac{1}{m_1} + \frac{1}{m_2} + \frac{1}{m_3} + m_1 m_2 m_3\right) - am_1 m_2 m_3\right\}^2\right]$$

$$= \frac{a^2}{4}\left[m_1^2 + m_2^2 + m_3^2 + \frac{1}{m_1^2} + \frac{1}{m_2^2} + \frac{1}{m_3^2} + m_1^2 + m_2^2 + m_3^2 + \frac{1}{m_1^2 m_2^2 m_3^2}\right]^2 \quad ...(5)$$

Comparing (4) and (5), we get $OA^2 = OD^2$ or $OA = OD$

Hence $OA = OB = OC = OD$. **Proved.**

Example 40(a):

The co-ordinates of the vertices of a triangle are (x_1, y_1), (x_2, y_2) and (x_3, y_3). The line joining the first two is divided in the ratio $l : k$, and the line joining this point of division to the opposite angular point is then divided in the ratio $m : k + l$. Find the co-ordinates of the, latter point of section.

Solution:

The co-ordinates of the vertices are given to be (x_1, y_1), (x_2, y_2) and (x_3, y_3). Let these points be A, B and C respectively. The co-ordinates of the point P which divides AB in the ratio l : k will be

$$\left(\frac{l\,x_2 + kx_1}{l+k}, \frac{l\,y_2 + ky_1}{l+k}\right)$$

The co-ordinates of the point Q which divides the join of P and C in the ratio m L : (k + l) will be

$$\left\{\frac{mx_3 + (k+l)\dfrac{l\,x_2 + kx_1}{(l+k)}}{m+k+l}, \frac{mx_3 + (k+l)\dfrac{l\,y_2 + ky_1}{(l+k)}}{m+k+l}\right\}$$

or $$\left(\frac{kx_1 + l\,x_2\; mx_3}{k+l+m}, \frac{ky_1 + l\,y_2\; my_3}{k+l+m}\right)$$ **Ans.**

Example 40(b):

Prove that the co-ordinates, x and y of the middle point of the line joining the point (2, 3) to the point (3, 4) satisfy the equation x – y +1 =0.

Solution:

The points are (2, 3) and (3, 4). The co-ordinates of the middle points joining the given points are

$$\left(\frac{2+3}{2}, \frac{3+4}{2}\right) \text{ or } \left(\frac{5}{2}, \frac{7}{2}\right)$$

The given equation is $x - y + 1 = 0$.

Substituting the co-ordinates of the mid-point in the given equation we get $\frac{5}{2} - \frac{7}{2} + 1 = 0$. Hence the co-ordinates satisfy the equation **Proved.**

Example 40(c):

Find the area of the triangle the co-ordinates of whose angular points are respectively ($a \cos \phi_1, b \sin \phi_1$). ($a \cos \phi_2, b \sin \phi_2$) and ($a \cos \phi_3, b \sin \phi_3$).

Solution:

The angular points are ($a \cos \phi_1, b \sin \phi_1$), ($a \cos \phi_2, b \sin \phi_2$) and ($a \cos \phi_3$, $b \sin \phi_3$).

Hence the area is

$$= \frac{1}{2}[(a\cos\phi_1 \times b\sin\phi_2 - b\sin\phi_1 \times a\cos\phi_2)$$

$$+ (a\cos\phi_2 \times b\sin\phi_3 - b\sin\phi_2 \times a\cos\phi_3)$$

$$+ (a\cos\phi_3 \times b\sin\phi_1 - b\sin\phi_3 \times a\cos\phi_1)]$$

$$= \frac{1}{2}ab[\cos\phi_1\sin\phi_2 - \sin\phi_1\cos\phi_2) + (\cos\phi_2\sin\phi_3 - \sin\phi_2\cos\phi_3)$$

$$+ (\cos\phi_3\sin\phi_1 - \sin\phi_3\cos\phi_1)]$$

$$= \frac{1}{2}ab[\sin(\phi_2 - \phi_1) + \sin(\phi_3 - \phi_2) + \sin(\phi_1 - \phi_3)]$$

$$= \frac{1}{2}ab\left[2\sin\frac{\phi_2 - \phi_1 + \phi_3 - \phi_2}{2}\cos\frac{\phi_2 - \phi_1 + \phi_3 + \phi_2}{2}\right.$$

$$\left. + 2\sin\left(\frac{\phi_1 - \phi_3}{2}\right)\cos\left(\frac{\phi_1 - \phi_3}{2}\right)\right]$$

$$= \frac{1}{2}ab\,2\sin\frac{\phi_3 - \phi_1}{2}\left[\cos\frac{2\phi_2 - \phi_3 - \phi_1}{2} - \cos\frac{\phi_1 - \phi_3}{2}\right]$$

$$\left\{\because \sin\frac{\phi_1 - \phi_3}{2} = -\sin\frac{\phi_3 - \phi_1}{2}\right\}$$

$$= ab.\sin\frac{\phi_3 - \phi_1}{2}.\,2\sin\frac{1}{2}\left(\frac{2\phi_2 - \phi_3 - \phi_1}{2} + \frac{\phi_1 - \phi_3}{2}\right)$$

$$\times \sin\frac{1}{2}\left(\frac{\phi_1 - \phi_3}{2} - \frac{2\phi_2 - \phi_3 - \phi_1}{2}\right)$$

$$= 2ab.\sin\frac{\phi_3 - \phi_1}{2}.\sin\frac{\phi_2 - \phi_3}{2}.\sin\frac{\phi_1 - \phi_2}{2}$$ **Ans.**

Example 41:

Prove that the points $(0, 0)\left(3, \frac{\pi}{2}\right)$ *and* $\left(3, \frac{\pi}{6}\right)$ *form an equilateral triangle.*

Solution:

Let the given points $(0, 0), \left(3, \frac{\pi}{2}\right)$ and $\left(3, \frac{\pi}{6}\right)$ be A, B and C respectively. Then

$$AB^2 = 0^2 + 3^2 - 2.0 \cos\left(\frac{\pi}{2} - 0\right) = 9 \text{ units}$$

$$AC^2 = 0^2 + 3^2 - 2.0 \cos\left(\frac{\pi}{6} - 0\right) = 9 \text{ units}$$

$$BC^2 = 9 + 9 - 18 \cos\frac{\pi}{3} = 9 \text{ units.}$$

Therefore AB = BC = CA. **Hence Proved.**

Formula:

If (r_1, θ_1), (r_2, θ_2), and (r_3, θ_3) be the angular points of a triangle, the area of that triangle is given by

$$= \frac{1}{2}\left[r_2\, r_1 \sin(\theta_2 - \theta_1) + r_3\, r_2 \sin(\theta_3 - \theta_2) + r_1\, r_3(\theta_1 - \theta_3)\right].$$

Example 42:

Find the area of the triangle the co-ordinates of whose angular points are (1, 30°), (2, 60°) and (3, 90°).

Solution:

The angular points are given to be (1, 30°), (2, 60°), (2, 90°). Hence area

$$= \frac{1}{2}[2.1. \sin(60° - 30°) + 3.2. \sin(90° - 60°) + 3.1. \sin(30° - 90°)]$$

$$= \frac{1}{2}[2 \sin 30° + 6 \sin 30° - 3 \sin 60°]$$

$$= \frac{1}{2}\left[4 - 3.\frac{\sqrt{3}}{2}\right] = \frac{1}{4}[8 - 3\sqrt{3}] \text{ units.}$$ **Ans.**

Example 43:

Transform to Cartesian co-ordinates the equation

$$r^2 = a^2 \cos 2\theta.$$

Solution:

The given equation is $r^2 = a^2 \cos 2\theta$.

or $r^2 = a^2 (\cos^2\theta - \sin^2\theta)$

or $r^2\, r^2 = a^2 (r^2 \cos^2\theta - r^2 \sin^2\theta)$ (multiplying both sides by r^2)

or $(x^2 + y^2)^2 = a^2 (x^2 - y^2)$ **Ans.**

Example 44(a):

Transform to Cartesian co-ordinates the equation

$$r^2 \sin 2\theta = 2a^2$$

Solution:

The given equation is $r^2 \sin 2\theta = 2a^2$.

or $r^2.2 \sin\theta \cos\theta = 2a^2$

or $r \sin\theta \, r\cos\theta = a^2$ or $yx = a^2$ or $xy = a^2$. **Ans.**

Example 44(b):

Transform to Cartesian co-ordinates the equation

$$r^2 \cos 2\theta = a^2$$

Solution:

The given equation is $r^2 \cos 2\theta = a^2$.

or $r^2 (\cos^2\theta - \sin^2\theta) = a^2$ or $r^2\cos^2\theta - r^2\sin^2\theta = a^2$

or $x^2 - y^2 = a^2$. **Ans.**

Example 44(c):

Find the Cartesian co-ordinates (drawing a figure) of the point whose polar co-ordinates are $\left(5, \frac{\pi}{4}\right)$.

Solution:

Point is $\left(5, \frac{\pi}{4}\right)$

$\therefore x = r\cos\theta = 5\cos\frac{\pi}{4} = \frac{5}{\sqrt{2}}$,

$y = r\sin\theta = 5\sin\frac{\pi}{4} = \frac{5}{\sqrt{2}}$.

Hence point is $\left(\frac{5}{\sqrt{2}}, \frac{5}{\sqrt{2}}\right)$.

Example 45:

Find the Cartesian co-ordinates (drawing a figure) of the point whose polar co-ordinates are $\left(-5, \frac{\pi}{3}\right)$.

Solution:

The point is $\left(-5, \frac{\pi}{3}\right)$

$$x = r\cos\theta = -5\cos\frac{\pi}{3} = \frac{5}{2},$$

$$y = r\sin\theta = -5\sin\frac{\pi}{3} = \frac{5\sqrt{3}}{2}.$$

Hence the point is $\left(\frac{-5}{2}, \frac{-5\sqrt{3}}{2}\right)$.

Example 46:

Transform to Cartesian co-ordinates the equation

$$r^{\frac{1}{2}}\cos\frac{\theta}{2} = a^{\frac{1}{2}}$$

Solution:

The given equation is $r^{\frac{1}{2}}\cos\frac{\theta}{2} = a^{\frac{1}{2}}$.

Squaring both sides, we get $r\cos^2\frac{\theta}{2} = a$

or $\quad r\left(\frac{1+\cos\theta}{2}\right) = a$

or $\quad r + r\cos\theta = 2a$ or $r = 2a - r\cos\theta$

or $\quad \sqrt{(x^2 + y^2)} = 2a - x$ or $x^2 + y^2 = (2a - x)^2$

or $\quad x^2 + y^2 = 4a^2 + x^2 - 4ax$ or $y^2 + 4ax = 4a^2$. **Ans.**

Example 47(a):

Find the Cartesian co-ordinates (drawing a figure) of the point whose polar co-ordinates are $\left(5, -\frac{\pi}{4}\right)$.

Solution:

The point is $\left(5, -\frac{\pi}{4}\right)$.

$$\therefore x = r\cos\theta = 5\cos\left(-\frac{\pi}{4}\right) = \frac{5}{\sqrt{2}}.$$

$$y = r\sin\theta = 5\sin\left(-\frac{\pi}{4}\right) = -\frac{5}{\sqrt{2}}.$$

Hence the point is $\left(\frac{-5}{2}, \frac{-5}{\sqrt{2}}\right)$

Example 47(b):

Change to polar co-ordinates the equation $x^2 + y^2 = a^2$

Solution:

The given equation $x^2 + y^2 = a^2$

or $\quad r^2 = a^2$

$\therefore \quad r = \sqrt{(x^2 + y^2)}$

or $\quad r = a.$ **Ans.**

Example 48:

Change to polar co-ordinates the equation $y = x \tan \alpha$.

Solution:

The given equation is $y = x \tan \alpha$.

$\Rightarrow \quad r \sin \theta = r \cos \theta$, than α $\quad\quad \therefore\ x = r \cos \theta.$

$\Rightarrow \quad \tan \theta = \tan \alpha$ $\quad\quad y = r \sin \theta.$

$\theta = \alpha.$ **Ans.**

Example 49:

Change to polar co-ordinates the equation $x^3 = y^2 (2a - x)$

Solution:

The given equation is $x^3 = y^2 (2a - x)$

$\Rightarrow \quad r^3 \cos^3 \theta = r^2 \sin^2 \theta (2a - r \cos \theta)$ $\quad\quad \therefore\ x = r \cos\theta.$

$\Rightarrow \quad r \cos^3 \theta = 2a \sin^2 \theta - r \sin \theta \cos \theta$ $\quad\quad x = r \sin\theta.$

$\Rightarrow \quad r \cos^3 \theta + r \sin^2 \theta \cos \theta = 2a \sin^2 \theta$

$\Rightarrow \quad r \cos \theta (\cos^2 \theta + \sin^2 \theta) = 2a \sin^2 \theta$

$r \cos \theta = 2a \sin^2 \theta$ **Ans.**

Example 50:

Change to polar co-ordinates the equation $(x^2 + y^2)^2 = a^2 (x^2 - y^2)$

Solution:

The given equation is $(x^2 + y^2)^2 = a^2 (x^2 - y^2)$

$\Rightarrow \quad (r^3)^2 = a^2 (r^2 \cos^2 \theta - r^2 \sin^2 \theta)$

$\Rightarrow \quad r^4 = a^2r^2 (\cos^2 \theta - \sin^2 \theta)$

$\Rightarrow \quad r^2 = a^2 \cos 2\theta$ **Ans.**

Formula:

The length of the line joining P and Q whose polar co-ordinates are (r_1, θ_1) and (r_2, θ_2) is given by

$$PQ^2 = r_1^{\ 2} + r_2^{\ 2} - 2r_1r_2 \cos(\theta_1, \theta_2).$$

Example 51(a):

Find the length of the straight line joining the paris of points whose polar co-ordinates are (2, 30°) and (4, 120°)

Solution:

The given points are (2, 30°) and (4, 120°). Hence the required length

$= \sqrt{\{2^2 + 4^2 - 2.2.4 \cos(30° = 120°)\}}$

$= \sqrt{(4 + 16 - 16 \cos 90)} = 2\sqrt{5}$ units. **Ans.**

Example 51(b):

Find the length of the straight line joining the paris of points whose polar co-ordinates are (–3, 45°) and (7, 105°)

Solution:

The given points are (–3, 45°) and (7, 105°). Hence the required distance

$= \sqrt{\{(3^2 + 7^2 - 2(-3)(7)\cos(45° - 105°)\}}$

$= \sqrt{(9 + 49 + 42 \cos 60°)} = \sqrt{79}$ units. **Ans.**

Example 51(c):

Find the length of the straight line joining the paris of points whose polar co-ordinates are $\left(a, \frac{\pi}{2}\right)$ and $\left(3a, \frac{\pi}{6}\right)$.

Solution:

The co-ordinates of the points are given to be

$$\left(a, \frac{\pi}{2}\right) \text{ and } \left(3a, \frac{\pi}{6}\right)$$

The required distance $= \sqrt{\left\{a^2 + 9a^2 - 2.a.3a. \cos\left(\frac{\pi}{2} - \frac{\pi}{6}\right)\right\}}$

$= \sqrt{\left(10a^2 - 6a^2 \cos\frac{\pi}{3}\right)} = a\sqrt{7}$ units. **Ans.**

Example 52:

Find the polar co-ordinates (drawing the figure) of the points $x = \sqrt{3}$, $y = 1$.

Solution:

$$x = \sqrt{3},\ y = 1$$

$$r = \sqrt{(3 + 1)} = 2$$

$$\theta = \tan^{-1}\left(\frac{y}{x}\right) = \tan^{-1}\left(\frac{1}{\sqrt{3}}\right)$$

$$\therefore\ \theta = \frac{\pi}{6}$$

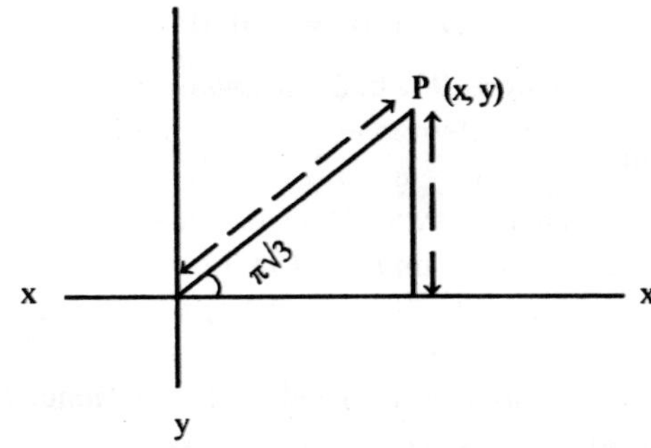

Hence point is $\left(2, \frac{\pi}{6}\right)$

Example 53:

Find the polar co-ordinates (drawing the figure) of the points $x = -\sqrt{3}$, $y = 1$.

Solution:

We have

$x = -\sqrt{3}$, $y = 1$. $\therefore\ r = \sqrt{(3 + 1)} = 2$

$$\theta = \tan^{-1}\left(\frac{y}{x}\right) = \tan^{-1}\left(-\frac{1}{\sqrt{3}}\right)$$

$\therefore\ \theta = \frac{5\pi}{6}$; $\therefore$ Point is $\left(2, \frac{5\pi}{6}\right)$

Example 54:

Find the polar co-ordinates (drawing the figure) of the points $x = -1$, $y = 1$.

Solution:

We have

$$x = -1,\ y = 1.$$

$$\therefore \quad r = \sqrt{(1 + 1)} = \sqrt{2}$$

$$\theta = \tan^{-1}\left(\frac{y}{x}\right) = \tan^{-1}\left(-\frac{-1}{1}\right)$$

$$\because \theta = \frac{3\pi}{4}; \quad \therefore \text{ Point is } \left(\sqrt{2}, \frac{3\pi}{4}\right)$$

Example 55:

Find the area of the triangle the co-ordinates of whose angular points are (–3, –30°), (5, 150°), and (7, 210°).

Solution:

The angular points are (–3, –30°), (5, 150°), and (7, 210°). So area of a triangle

$$\Delta = \frac{1}{2}\ [5.\ (-3) \sin (150° + 30°) + 7.5 \sin (210° + 150°) + (-3)\ (7) \sin$$
$$(-30° - 210°)$$

$$= \frac{1}{2}\ [-15 \sin 180° + 35 \sin 60° + 21 \sin 240°]$$

$$= \frac{1}{2} = \frac{1}{2}\left[35.\frac{\sqrt{3}}{2} - 21.\frac{\sqrt{3}}{2}\right] = \frac{7}{2}\sqrt{3} \text{ units.}$$ **Ans.**

Example 56:

Find the area of the triangle the co-ordinates of whose angular points are

$$\left(-a, \frac{\pi}{6}\right), \left(a, \frac{\pi}{2}\right), \text{ and } \left(-2a, \frac{2\pi}{3}\right).$$

Solution:

The angular points are $\left(-a, \frac{\pi}{6}\right), \left(a, \frac{\pi}{2}\right)$, and $\left(-2a, \frac{2\pi}{3}\right)$

Hence the area of a triangle

$$\Delta = \frac{1}{2}\left[a.(-a)\sin\left(\frac{\pi}{2}-\frac{\pi}{6}\right) + (-2a)(a)\sin\left(-\frac{2\pi}{3}-\frac{\pi}{2}\right)\right.$$
$$\left. + (-a)(-2a)\sin\left(\frac{\pi}{6}+\frac{2\pi}{3}\right)\right]$$

$$= \frac{1}{2}\left[-a^2\sin\frac{\pi}{3} + 2a^2\sin\frac{7\pi}{6} + 2a^2\sin\frac{5\pi}{6}\right]$$

$$= \frac{1}{2}\left[-a^2.\frac{\sqrt{3}}{2} - 2a^2\frac{1}{2} + 2a^2\frac{1}{2}\right] = \frac{a^2\sqrt{3}}{2}$$ units (omitting –ve sign) **Ans.**

Working Rule: To change polar co-ordinates into Cartesian co-ordinates, put $x \equiv r\cos\theta$ and $y \equiv r\sin\theta$ and to change the cartesian co-ordinates in polar co-ordinates.

Put $r = \sqrt{(x^2 + y^2)}$

and $\theta = \tan^{-1}\left(\frac{y}{x}\right)$

Example 57:

Prove (by showing that the area of the triangle formed by them is zero) that the following sets of three points are in a straight line : (1, 4) (3, 2) and (3, 16).

Solution:

The given points are (1, 4), (3, –2) and (–3, 16).

$$= \frac{1}{2}\left[\{1\times(-2) - 4.3\} + \{3.16 - (-2)(-3)\} + \{(-3)\times 4 - (16\times 1)\}\right]$$

$$= \frac{1}{2}\left[-2 - 12 + 48 - 6 - 12 - 16\right] = 0$$ **Hence proved.**

Example 58:

Prove (by showing that the area of the triangle formed by them is zero) that the following sets of three points are in a straight line : (–1/2, 3) (–5, 6) and (–8, 8).

Solution:

The points are given to be (–1/2, 3) (–5, 6) and (–8, 8).

The area enclosed by them is given as

$$= \frac{1}{2}\left[\left\{\left(-\frac{1}{2}\right) \times 6 - 3(-5) + (-5) \times 8 - 6(-8) + (-8) \times 3 - 8\left(-\frac{1}{2}\right)\right\}\right]$$

$$= \frac{1}{2}\ [-3 = 15 - 40 + 48 - 24 + 4] = 0$$ **Hence proved.**

Example 59:

Prove (by showing that the area of the triangle formed by them is zero) that the following sets of three points are in a straight line : (a, b + c), (b, c + a), and (c, a + b).

Solution:

The angulár points are (a, b + c), (b, c + a), and (c, a + b).

The area enclosed by them is given by

$$= \frac{1}{2}[\{a\,(c + a) - b\,(b + c)\} + b\,(a + b) - c\,(a + c) + c\,(b + c) - a\,(a + b)\}]$$

$$= \frac{1}{2}\ [ac + a^2 - b^2 - bc + ba + b^2 - c^2 - ca + cb + c^2 - a^2 - ab]$$

$= 0$ **Hence Proved.**

Example 60:

Find the area of the quadrilateral the co-ordinates of whose angular points, taken in order, are (1, 1), (3, 4), (5, –2) and (4, –7).

Solution:

The angular points of the quadrilateral all taken in order are (1, 1), (3, 4), (5, –2) and (4, –7).

Applying the formula for the area of the polygon, we get area

$$= \frac{1}{2}\ [\{(1 \times 4) - 1 \times 3)\} + \{3 \times (-2) - 4 \times 5\} + \{5\,(-7) - (-2)\,(4)\} + \{4 \times 1 - 1\,(-7)\}]$$

$$= \frac{1}{2}\ [4 - 3 - 6 - 20 - 35 + 8 + 4 + 7]$$

$$= \frac{41}{2} = 20\frac{1}{2}$$ (omitting the +ve sign). **Ans.**

Example 61:

Find the area of the quadrilateral the co-ordinates of whose angular points, taken in order, are (–1, 6), (–3, –9), (5, –8) and (3, 9).

Solution:

The co-ordinates of the angular points of the quadrilateral taken in order are (–1, 6), (–3, –9), (5, –8) and (3, 9).

Hence the area of the quad. is given by

$$= \frac{1}{2}[\{(-1) \times (-9) - (6)(-3)\} + \{(-3)(-8) - (-9) - (5)\} + \{(5)(9) - (-8)(3)\} + [(3)(6) - (9)(-1)\}]$$

$$= \frac{1}{2}[9 + 18 + 24 + 45 + 45 + 24 + 18 + 9]$$

= 96 units. **Ans.**

Example 62:

Find the area of the triangle the co-ordinates of whose angular points are respectively $(am_1^2, 2am_1)$, $(am_2^2, 2am_2)$ and $(am_3^2, 2am_3)$

Solution:

The angular points are $(am_1^2, 2am_1)$, $(am_2^2, 2am_2)$ and $(am_3^2, 2am_3)$

Hence the area of the triangle

$$\Delta = \frac{1}{2}\left[\left(am_1^2 \times 2am_2 - 2am_1 \times am_2^2\right) + \left(am_2^2 \times 2am_3 - 2am_2 \times am_3^2\right) + \left(am_3^2 \times 2am_1 - 2am_3 \times am_1^2\right)\right]$$

$$= \frac{1}{2}2a^2\left[m_1^2\, m_2 - m_1\, m_2^2 + am_2^2\, m_3 - m_2\, am_3^2 + m_3^2\, m_1\, m_3\, m_1^2\right]$$

Rearranging the terms according to the powers of m_1, we get

$= a^2\, [m_1^2m_2 - m_3m_1^2 - m_1m_2 + m_3^2m_1 + m_2^2m_3 - m_2m_3^2]$

$= a^2\, [m_1^2\,(m_2 - m_3) - m_1\,(m_2^2 - m_3^2) + m_2m_3\,(m_2 - m_3)]$

$= a^2\,(m_2 - m_3)\,[m_1^2 - m_1\,(m_2 + m_3) + m_2m_3]$

$= a^2\,(m_2 - m_3)\,[m_1^2 - m_1m_2 - m_1m_3 + m_2m_3]$

$= a^2\,(m_2 - m_3)\,[m_1\,(m_1 - m_2) - m_3\,(m_1 - m_2)]$

$= a^2\,(m_2 - m_3)\,(m_3 - m_1)\,(m_1 - m_2)$ (omitting –ve sign) **Ans.**

Example 63:

If O be the origin, and if the co-ordinates of any two points P_1 and P_2 be respectively (x_1, y_1) and (x_2, x_2), prove that OP_1, OP_2, cos P_1 $PO_2 = x_1x_2 + y_1y_2$.

Solution:

By simple trigonometry, we have

$P_1P_2^2 = PO_1^2 + OP_2^2 - 2OP_1\, OP_2 \times \cos P_1\, OP_2.$

Straight lines and areas of triangles

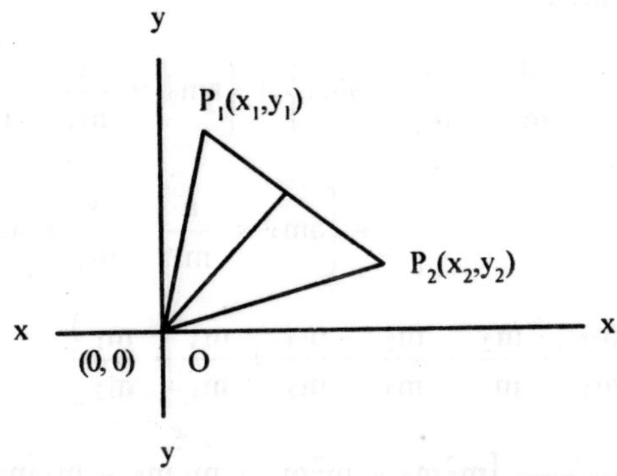

Hence $OP_1\, OP_2 \cos P_1\, OP_2 = \frac{1}{2}\left[OP_1^2 + OP_2^2 - P_1P_2^2\right]$ ...(1)

Now $OP_1^2 = (x_1 - 0)^2 + (y_1 - 0)^2 = x_1^2 + y_1^2,$

$OP_2^2 = (x_2 - 0)^2 + (y_2 - 0)^2 = x_2^2 + y_2^2,$

$P_1P_2^2 = (x_1 - x_2)^2 + (y_1 - y_2)^2$

$= x_1^2 + x_2^2 - 2x_1x_2 + y_1^2 + y_2^2 - 2y_1y_2$

Substituting the values in equation No. (1), we get

$$OP_1,\ OP_2.\ \cos P_1\, OP_2 = \frac{1}{2}\left[\left(x_1^2 + y_1^2\right) + \left(x_2^2 + y_2^2\right) - \left(x_1^2 + x_2^2 - 2x_1x_2 + y_1^2 + y_2^2 - 2y_1y_2\right)\right]$$

$= x_1x_2 + y_1y_2.$ **Proved.**

Example 64:

Find the area of the triangle the co-ordinates of whose angular is are respectively

$$\left\{am_1, \frac{a}{m_1}\right\}, \left\{am_2, \frac{a}{m_2}\right\} \text{ and } \left\{am_3, \frac{a}{m_3}\right\}$$

Solution:

The angular points are given to be

$$\left\{am_1, \frac{a}{m_1}\right\}, \left\{am_2, \frac{a}{m_2}\right\} \text{ and } \left\{am_3, \frac{a}{m_3}\right\}$$

Hence the area

$$= \frac{1}{2}\left[\left\{am_1 \times \frac{a}{m_2} - \frac{a}{m_1} \times am_2\right\} + \left\{am_2 \times \frac{a}{m_3} - \frac{a}{m_2} \times am_3\right\} + \left\{am_3 \times \frac{a}{m_1} - \frac{a}{m_3} \times am_1\right\}\right]$$

$$= \frac{1}{2} a^2 \left[\frac{m_1}{m_2} - \frac{m_2}{m_1} + \frac{m_2}{m_3} - \frac{m_3}{m_2} + \frac{m_3}{m_1} - \frac{m_1}{m_3}\right]$$

$$= \frac{1}{2} a^2 \frac{1}{m_1 m_2 m_3}\left[m_1^2 m_3 - m_2^2 m_3 + m_1 m_2^2 - m_1 m_3^2 + m_2 m_3^2 - m_1^2 m_2\right]$$

Taking -ve sign common and rearranging, we get

$$= \frac{1}{2} \frac{a^2}{m_1 m_2 m_3}\left[m_1^2 m_2 - m_1^2 m_3 - m_1 m_2^2 + m_1 m_3^2 + m_2^2 m_3 - m_2 m_3^2\right]$$

$= a^2 (m_2 - m_3) (m_3 - m_1) (m_1 - m_2)/2m1m_2m3$

(as in Q No. 7 of this example). **Ans.**

Article:

If three points be in a line, the area of the triangle formed by joining these points will be zero and, conversely; if the area of the triangle formed by joining 3 points be zero, the given points are in a straight line.

Example 65:

Find the area of the triangle the co-ordinates of whose angular points are respectively (a, b + c), (a, b – c) and (–a, c).

Solution:

The angular points are (a, b + c), (a, b – c) and (–a, c).

Hence the area

$$= \frac{1}{2}[\{a(b-c) - a(b+c)\} + \{a.c - (b-c)(-a)\}] + \{(-a)(b+c) - c\,a\}]$$

$$= \frac{1}{2}[ab - ac - ab - ac + ac + ab - ac - ab - ac - ac] = 2ac$$

(omitting the negative sing.) **Ans.**

Example 66(a):

Find the area of the triangle the co-ordinates of whose angular points are respectively (a, c + a), (a, c) and (–a, c – a).

Solution:

The angular points are (a, c + a), (a, c) and (–a, c – a).

Hence area of triangle

$$\Delta = \frac{1}{2}[a.c - a(c+a)\} + \{a(c-a) - c(-a)\} + \{(-a)(c+a) - (c-a)$$

$$- (c-a).a\}]$$

$$= \frac{1}{2}[ab - ac - a^2 + ac - a^2 + ac - ac - a^2 - ac + a^2]$$

$= a^2$ (omitting the negative sing.) **Ans.**

Example 66(b):

Find the area of the triangle the co-ordinates of whose angular points are respectively (1, 3), (–7, 6) and (5, –1).

Solution:

Co-ordinates of the angular points are (1, 3), (–7, 6) and (5, –1); hence

$$\text{Area} = \frac{1}{2}[(1.6 - 3(-7)\} + \{(-7)(-1) - 6.5\} + \{5.3 - (-1)\,1)\}]$$

$$= \frac{1}{2}[6 + 21 + 7 - 30 + 15 + 1]$$

= 10 units. **Ans.**

Example 66(c):

Find the area of the triangle the co-ordinates of whose angular points are respectively (0, 4), (3, 6) and (–8, –2).

Solution:

The angular points are (0, 4), (3, 6) and (–8, –2).

Hence the area is

$$= \frac{1}{2}\ [0.6 - 4.3) + 3.(-2) - 6.(-8)\} + \{(-8).4 - (-2).0\}]$$

$$= \frac{1}{2}\ [0 - 12 - 6 + 48 + 32 - 0] = 1 \text{ unit (omitting the –ve sign)}$$ **Ans.**

Example 67:

Find the area of the triangle the co-ordinates of whose angular points are respectively (5, 2), (–9, –3) and (–3, –5).

Solution:

The angular points are (5, 2), (–9, –3) and (–3, –5).

Hence area

$$= \frac{1}{2}\ [\{5\,(-3) - (2)(-9)\} + \{(-9)(-5) - (-3)(-3)\} + \{(-3)\,2 - (-5)\,5)\}]$$

$$= \frac{1}{2}\ [-15 + 18 + 45 - 96 + 25] = 29 \text{ unit}$$ **Ans.**

Example 68:

Find the area of the triangle the co-ordinates of whose angular points are respectively $\{am_1m_2,\ a\,(m_1 + m_2)\}$ $\{am_2m_3,\ a\,(m_2 + m_3)\}$ and $\{am_3m_1,\ a\,(m_3 + m_1)\}$.

Solution:

The angular points are $\{am_1m_2, a\,(m_1 + m_2)\}$ $\{am_2m_3, a\,(m_2 + m_3)\}$ and $\{am_3m_1, a\,(m_3 + m_1)\}$.

Hence the area is

$$= \frac{1}{2}\ [\{am_1m_2 \times a\,(m_2 + m_3) - a\,(m_1 + m_2) \times am_2\,m_3\} + \{am_2m_3 \times a\,(m_3 + m_1) - a\,(m_2 + m_1) \times am_3m_1\} + \{am_3\,m_1 \times a\,(m_1{+}m_2) - a\,(m_3 + m_1) \times am_1m_2\}]$$

$$= \frac{1}{2}\,a^2\left[m_1m_2^2 + m_1m_2m_3 - m_1m_2m_3 - m_2^2m_3 + m_2\,m_3^2 + m_1m_2m_3 - m_1m_2m_3 - m_3m_1^2 + m_1^2m_3 + m_3m_2m_1 - m_3m_2m_1 - m_1^2m_2\right]$$

$$= \frac{1}{2} a^2 \left[m_1 m_2^2 - m_2^2 m_3 + m_2 m_3^2 - m_3^2 m_1 + m_3 m_1^2 - m_1^2 m_2\right]$$

Taking –ve sign common and rearranging, we get

$$= \frac{1}{2} a^2 \left[m_1^2 m_2 - m_1^2 m_3 + m_1 m_2^2 - m_1 m_3^2 + m_2^2 m_3 - m_2 m_3^2\right]$$

$$= \frac{1}{2} a^2 (m_2 - m_3)(m_3 - m_1)(m_1 - m_2)$$ (as in previous equation). **Ans.**

Example 69(a):

Transform to Cartesian co-ordinates the equation

$$r^{1/2} = a^{1/2} \sin \frac{\theta}{2}$$

Solution:

The given equation is $r^{1/2} = a^{1/2} \sin \frac{\theta}{2}$

Squaring both sides, we get $r = a \sin \frac{\theta}{2}$

$$r = a\left(\frac{1 - \cos\theta}{2}\right)$$

or $2r = a = a \cos\theta$ or $2r^2 = ar - ar \cos\theta$ (multiplying both sides by r)

or $2(x^2 + y^2) = a\sqrt{(x^2 + y^2)} - a.x$ or $2x^2 + 2y^2 + ax = a\sqrt{(x^2 + y^2)}$

or $4x^4 + 4y^4 + a^2x^2 + 8x^2y^2 + 4x^2ax + 4y^2ax = a\,(x^2 + y^2)$

(on squaring both sides)

or $4(x^4 + y^4 + 2x^2y^2 + x^2.ax + y^2\,ax) = a^2y^2$

or $4[(x^2 + y^2) + ax\,(x^2 + y^2) = a^2y^2$

or $4(x^2 + y^2)\,(x^2 + y^2 + ax) = a^2y^2$ **Ans.**

Example 69(b):

Transform to Cartesian co-ordinates the equation r (cos 3θ + sin 3θ) = 5k sin θ cos θ.

Solution:

The given equation is

$$r\,(\cos 3\theta + \sin 3\theta) = 5k \sin\theta \cos\theta$$

or $r\,(4\cos^3\theta - 3\cos\theta + 3\sin\theta - 4\sin^3\theta) = 5k \sin\theta \cos\theta$

or $4r\,(\cos^3\theta - \sin^3\theta) - 3r\,(\cos\theta - \sin\theta) = 5k \sin\theta \cos\theta$

Multiplying both sides by r^2, we get

$$4(r^3 \cos^3 \theta - r^3 \sin^3 \theta) - 3r^2(r \cos \theta - r \sin \theta) = 5k\ r \sin \theta.\ r \cos\theta$$

or $4(x^3 - y^3) - 3\ (x^2 - y^2)\ (x - y) = 5k.yx$

or $4x^3 - 4y^3 - 3x^3 + 3x^3y - 3y^2x + 3y^3 = 5kyx$

or $x^3 + 3x^2y - 3xy^3 + y^3 = 5kyx$ **Ans.**

Example 70:

Transform to Cartesian co-ordinates the equation

$$\theta = \tan^{-1} m.$$

Solution:

The given equation is $\theta = \tan^{-1} m$ or $\tan \theta = m$ ...(1)

and $\therefore$ $y = r \sin \theta$ and $x = r \cos \theta$, $\therefore$ $y/x = \tan \theta$

Putting in (1), we get $y/x = m$ or $y = mx$. **Ans.**

Example 71:

Transform to Cartesin co-ordinates the equation $r = a \cos \theta$

Solution:

The given equation is $r = a \cos \theta$

$\Rightarrow r^2 = a.r \cos \theta$ (multiplying both sides by r)

$\Rightarrow x^2 + y^2 = a.x$ **Ans.**

Example 72(a):

Transform to Cartesian co-ordinates the equation $r = a \sin 2\theta$.

Solution:

The given equation is $r = a \sin \theta$

$\Rightarrow r = a.2 \sin \theta \cos \theta$

$\Rightarrow r.r^2 = 2a.r. \sin \theta.\ r \cos \theta$ (multiplying both sides by r^2)

$\Rightarrow (x^2 + y^2)^{3/2} = 2a.yx$

$\Rightarrow (x^2 + y^2)^3 = 4a^2x^2y^2$. **Ans.**

Example 72(b):

Find the equations to the sides of the triangle the co-ordinates of whose angular points are respectively (1, 4) (2, 3) and (–1, –2).

Solution:

The points are given to be (1, 4) (2, 3) and (–1, –2).

Let these points be A, B and C respectively; then the equation of AB is

$$y - 4 = \frac{-3-4}{2-1}(x - 1) \text{ or } y + 7x = 11. \qquad \textbf{Ans.}$$

Equation of BC is $y - (-3) = \dfrac{-2-(-3)}{-1-2}$ $(x - 2)$ or $x + 3y + 7 = 0$. **Ans.**

Equation of CA is $y - 4 = \dfrac{-2-4}{-1-1}$ $(x - 1)$ or $y - 3x = 1$.

Example 72(c):

Find the equations to the sides of the triangle the co-ordinates of whose angular points are respectively (0, 1) (2, 0) and (–1, –2).

Solution:

The co-ordinates are given to be (0, 1) (2, 0) and (–1, –2).

Let these points be A, B and C respectively; then the equation of AB is

$$y - 1 = \frac{0-1}{2-0}(x - 0) \text{ or } x + 2y = 2. \qquad \textbf{Ans.}$$

The equation of BC will be $y - 0 = \dfrac{-2-0}{-1-2}$ $(x - 2)$ or $2x - 3y = 4$. **Ans.**

The equation of CA will be $y - 1 = \dfrac{-2-1}{-1-0}$ $(x - 0)$ or $y - 3x = 1$. **Ans.**

Example 73:

Find the equation to the straight line passing through the pairs of points ($a \cos \phi_1$, $b \sin \phi_1$) and ($a \cos \phi_2$, $b \sin \phi_2$).

Solution:

The points are ($a \cos \phi_1$, $b \sin \phi_1$) and ($a \cos \phi_2$, $b \sin \phi_2$). Hence the equation is

$$y - b \sin \phi_1 = \frac{b \sin \phi_2 - b \sin \phi_1}{a \cos \phi_2 - a \cos \phi_1}(x - a \cos \phi_1)$$

$$\Rightarrow y - b \sin \phi_1 = \frac{b.2 \cos \dfrac{\phi_2 + \phi_1}{2} . \sin \dfrac{\phi_2 - \phi_1}{2}}{-a.2 \sin \dfrac{\phi_2 + \phi_1}{2} . \sin \dfrac{\phi_2 + \phi_1}{2}}(x - a \cos \phi_1)$$

$$\Rightarrow \text{ay sin}\frac{1}{2}(\phi_2+\phi_1) - \text{ba sin}\,\phi_1\,\text{sin}\frac{1}{2}(\phi_2+\phi_1) - \text{bx}\quad\cos\frac{1}{2}$$

$$(\phi_2+\phi_1) + \text{ab}\cos\phi_1\cos\frac{1}{2}(\phi_1+\phi_2)$$

$$\Rightarrow \text{ay sin}\frac{1}{2}(\phi_2+\phi_1) + \text{bx}\cos\frac{1}{2}(\phi_2+\phi_1) = \text{ab}\left[\cos\phi_1\cos\frac{1}{2}(\phi_2+\phi_1) + \sin\phi_1\sin\frac{1}{2}(\phi_2+\phi_1)\right]$$

$$\Rightarrow \text{bx}\cos\frac{1}{2}(\phi_2+\phi_1) + \text{ay}\sin\frac{1}{2}(\phi_2+\phi_1) = \text{ab}\cos\left(\phi_1 - \frac{\phi_2+\phi_1}{2}\right)$$

$$\Rightarrow \quad \frac{x}{a}\cos\frac{1}{2}(\phi_2+\phi_1) + \frac{y}{b}\sin\frac{1}{2}(\phi_2+\phi_1) = \cos\frac{1}{2}(\phi_1+\phi_2).\ \textbf{Ans.}$$

Example 74(a):

Find the equation to the straight line passing through the pairs of points (a sec ϕ_1, b tan ϕ_1) and (a sec ϕ_2, b tan ϕ_2).

Solution:

The points are (a sec ϕ_1, b tan ϕ_1) and (a sec ϕ_2, b tan ϕ_2). Hence the equation is

$$y - b\tan\phi_1 = \frac{b\tan\phi_2 - b\tan\phi_1}{a\sec\phi_2 - a\sec\phi_1}(x - a\sec\phi_1)$$

$$\Rightarrow \quad y - b\tan\phi_1 = \frac{b\left[\left(\frac{\sin\phi_2}{\cos\phi_2}\right) - \left(\frac{\sin\phi_1}{\cos\phi_1}\right)\right]}{a\left[\left(\frac{1}{\cos\phi_2}\right) - \left(\frac{1}{\cos\phi_1}\right)\right]}(x - a\sec\phi_1)$$

$$\Rightarrow y - \frac{b\tan\phi_1}{\cos\phi_1} = \frac{b[\sin\phi_2\cos\phi_2 - \sin\phi_1\cos\phi_1]}{a[\cos\phi_2 - \cos\phi_2]}\left(x - a\frac{1}{\cos\phi_1}\right)$$

$$\Rightarrow y - \frac{b\sin\phi_1}{\cos\phi_1} = \frac{b\sin(\phi_2 - \phi_1)}{a(\cos\phi_2 - \cos\phi_2)}\left(x - \frac{a}{\cos\phi_1}\right)$$

$$\Rightarrow y - \frac{b\sin\phi_1}{\cos\phi_1} = \frac{b.2\sin\left(\frac{\phi_2-\phi_1}{2}\right)\cos\left(\frac{\phi_2-\phi_1}{2}\right)}{a.2\sin\left(\frac{\phi_1-\phi_2}{2}\right)\sin\left(\frac{\phi_2-\phi_1}{2}\right)}\left(x - \frac{a}{\cos\phi_1}\right)$$

$$\Rightarrow y-\frac{b\sin\phi_1}{\cos\phi_1}=\frac{b\cos\left(\frac{\phi_1-\phi_2}{2}\right)}{a\sin\left(\frac{\phi_1-\phi_2}{2}\right)}\left(x-\frac{a}{\cos\phi_1}\right)$$

$$\Rightarrow ay\sin\left(\frac{\phi_1-\phi_2}{2}\right)-ab\frac{\sin\phi_1}{\cos\phi_1}\sin\left(\frac{\phi_1+\phi_2}{2}\right)$$

$$=bx\cos\frac{\phi_1-\phi_2}{2}-\frac{ab}{\cos\phi_1}\cos\left(\frac{\phi_1+\phi_2}{2}\right)$$

$$\Rightarrow bx\cos\frac{1}{2}(\phi_1-\phi_2)-ay\sin\frac{1}{2}(\phi_1-\phi_2)$$

$$\Rightarrow ab\left[\frac{1}{\cos\phi_1}\cos\left(\frac{\phi_1-\phi_2}{2}\right)-\frac{\sin\phi_1}{\cos\phi_1}\sin\left(\frac{\phi_1+\phi_2}{2}\right)\right]$$

$$\Rightarrow \frac{ab}{2\cos\phi_1}\left[2\cos\left(\frac{\phi_1-\phi_2}{2}\right)-2\sin\phi_1\sin\left(\frac{\phi_1+\phi_2}{2}\right)\right]$$

$$\Rightarrow \frac{ab}{2\cos\phi_1}\left[2\cos\left(\frac{\phi_1-\phi_2}{2}\right)-\left(\cos\frac{\phi_1+\phi_2}{2}-\cos\frac{3\phi_1+\phi_2}{2}\right)\right]$$

$$\Rightarrow \frac{ab}{2\cos\phi_1}\left[\cos\left(\frac{\phi_1-\phi_2}{2}\right)+\cos\left(\frac{3\phi_1+\phi_2}{2}\right)\right]$$

$$\Rightarrow \frac{ab}{2\cos\phi_1}\left[2\cos\phi_1\cos\left(\frac{\phi_1-\phi_2}{2}\right)\right]=ab\cos\left(\frac{\phi_1+\phi_2}{2}\right)$$

Hence the required equation is

$$bx\cos\frac{1}{2}=(\phi_1-\phi_2)-ay\sin\frac{1}{2}=(\phi_1+\phi_2)=ab\cos\frac{1}{2}(\phi_1+\phi_2)$$ **Ans.**

Example 74(b):

Find the equations to the straight lines which pass through the point (1, –2) and cut off equal distances from the two axes.

Solution:

Two cases are possible.

(i) Intercepts are equal, having the same sign. So let each intercept be a. Therefore the equation of the line will be $x/a + y/a = 1$.

The given point (1, –2) will satisfy it. So substituting the value of x and y, we get $\frac{1}{a} - \frac{2}{a} = 1$ or a = –1.

Hence the required equation is $\frac{x}{-1} + \frac{y}{-1} = 1$

or $$x + y + 1 = 0$$ **Ans.**

(ii) As the point (1, –2) is in the 4th quadrant, the x-intercept will be +ve and y-intercept will be negative. If the length of the intercepts be b, the equation becomes $\frac{x}{b} - \frac{y}{b} = 1$.

Substituting the co-ordinates of the point, we get

$$\frac{1}{b} - \frac{-2}{b} = 1 \quad \text{or} = 3$$

Hence the required equation is $\frac{x}{3} - \frac{y}{3} = 1$ or x – y = 3 **Ans.**

Example 75:

Find the equations to the straight lines which go through the origin and trisect the portion of the straight line 3x + y = 12 which is ***intercepted*** between the axes of co-ordinates.

Solution:

The given line is,

$$3x + y = 12 \quad \text{or} \quad \frac{x}{4} + \frac{y}{12} = 1.$$

Let the line cut the axes at A and B; then their co-ordinates will be (4, 0) and (0, 12) respectively as OA = 4 and OB = 12.

Let P and Q be the points of trisection. So P divides AB in the ratio of 1 : 2. The co-ordinates of P will be

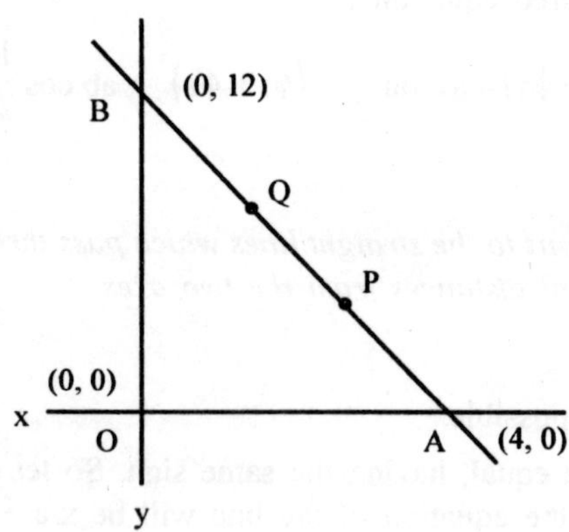

$$\left(\frac{1.0+2.4}{1+2}, \frac{1.12+2.0}{1+2}\right) \text{ or } \left(\frac{8}{3}, 4\right)$$

Point O is (0, 0). Hence the equation of OP is

$$y-0=\frac{4-0}{\frac{8}{3}-0}(x-0) \Rightarrow 2y = 3x$$ **Ans.**

The other point Q will divide AB in the ratio of 2 : 1 ; so the co-ordinates of Q are $\left(\frac{2.0+1.4}{2+1}, \frac{2.12+1.0}{2+1}\right)$ or $\left(\frac{4}{3}, 8\right)$.

Hence the equation of OQ is $y-0=\frac{8-0}{\frac{4}{3}-0}(x = 0)$ or y = 6x **Ans.**

Example 76:

Find the equation to the straight line which passes through the point (5, 6) and has intercepts on the axes.

1. equal in magnitude and both positive.
2. equal in magnitude but opposite in sign.

Solution:

(i) Let the length of intercept be a in both cases; then the equation will be x/a + y/a = 1.

Since it passes through the point (5, 6), these co-ordinates must satisfy the equation. Hence $\frac{5}{a}+\frac{6}{a}=1$ or a = 11.

∴ The required equation is $\frac{x}{11}+\frac{y}{11}=1$ or x + y = 11. **Ans.**

(ii) Let the intercept on x-axis be a, the intercept on y-axis will be –a and the equation of the line will be $\frac{x}{a}-\frac{y}{a}=1$

As it passes through (5, 6), the co-ordinates must satisfy the equation. Hence $\frac{5}{a}-\frac{6}{a}=1$ or a = –1.

∴ The required equation is $\frac{x}{-1}-\frac{y}{-1}=1$

⇒ y – x = 1. **Ans.**

Example 77:

Find the equation to the straight line which passes through the given point (x', y') and is such that the given point bisects the part intercepted between the axes.

Solution:

Let the straight line cuts the axes of x and y at A (a, 0) and B (0, b) respectively. The co-ordinates of the mid-point of AB, i.e. P will be

$$\left(\frac{a+0}{2}, \frac{0+b}{2}\right)$$

The co-ordinates of P are given to be (x', y').

Hence $\quad x' = \dfrac{a}{2}$ and $y' = \dfrac{b}{2}$

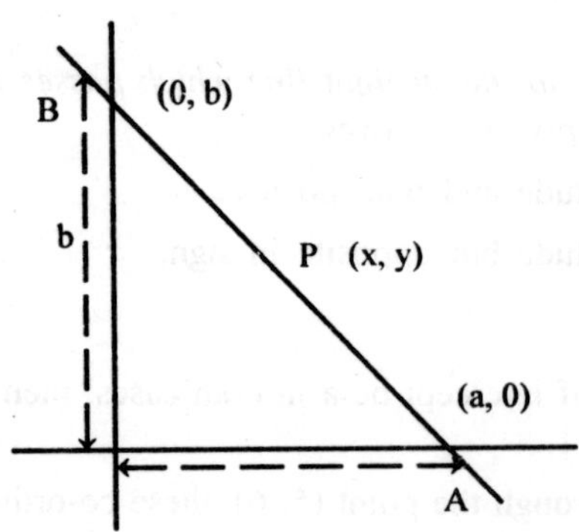

$\Rightarrow \qquad 2x' = a$ and $2y' = b$

Hence the required equation is $\dfrac{x}{2x'} + \dfrac{y}{2y'} = 1$ or $xy' + yx' = 2x'y'$.

Example 78:

Find the equation to the straight line which passes through the point (–4, 3) and is such that the portion of it between the axes is divided by the point in the ratio 5 : 3.

Solution:

Let the required straight line cuts the axes of x and y at A (a, 0) and B (0, b) respectively. Hence the co-ordinates of the point P which divides AB in the ratio of 5 : 3 are given by

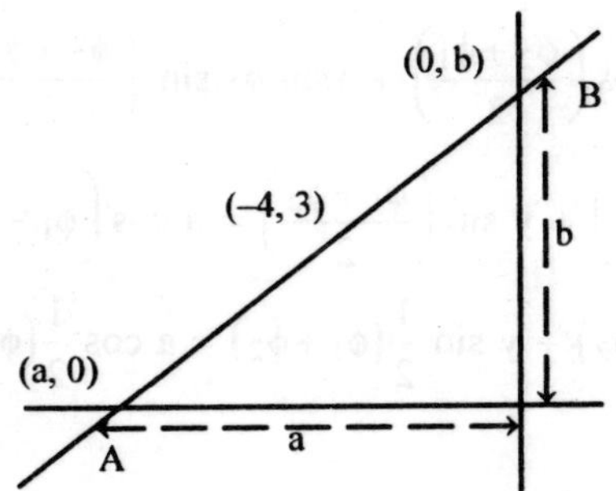

$$x = \frac{3a}{8} \text{ and } y = \frac{5b}{8}$$

By hypothesis, co-ordinates of P are (–4, 3).

Hence $x = \frac{3a}{8} = -4$ and $\frac{5b}{8} = 3$ or $a = \frac{32}{-3}$ and $b = \frac{24}{5}$.

$\therefore$ The required equation is $\dfrac{x}{-\frac{32}{3}} + \dfrac{y}{\frac{24}{5}} = 1$

$\Rightarrow \quad 20y - 9x = 96.$ **Ans.**

Example 79:

Find the equation to the straight line passing through the pairs of points $(a \cos \phi_1, a \sin \phi_1)$ and $(a \cos \phi_2, a \sin \phi_2)$

Solution:

Points are $(a \cos \phi_1, a \sin \phi_1)$ and $(a \cos \phi_2, a \sin \phi_2)$.
Hence the equation is

$$y - a \sin\phi_1 = \frac{a \sin\phi_2 - a \sin\phi_1}{a \cos\phi_2 - a \cos\phi_1}(x - a\cos\phi_1)$$

$$\Rightarrow y - a \sin\phi_1 = \frac{a \cos\frac{\phi_2 + \phi_1}{2} \sin\frac{\phi_2 + \phi_1}{2}}{-a.2 \sin\frac{\phi_2 + \phi_1}{2} \sin\frac{\phi_2 + \phi_1}{2}}(x - a\cos\phi_1)$$

$$\Rightarrow y \sin\frac{\phi_2 + \phi_1}{2} - a \sin\phi_1 \sin\frac{\phi_2 + \phi_1}{2} = x \cos\frac{\phi_2 + \phi_1}{2} + a\cos\phi_1 \cos\frac{\phi_2 + \phi_1}{2}$$

$$\Rightarrow y \sin\left(\frac{\phi_2 + \phi_1}{2}\right) + x \cos\left(\frac{\phi_2 + \phi_1}{2}\right)$$

$$= a\left[\cos\phi_1 \cos\left(\frac{\phi_2+\phi_1}{2}\right) + a\sin\phi_1 \sin\left(\frac{\phi_2+\phi_1}{2}\right)\right]$$

$$\Rightarrow y\cos\left(\frac{\phi_1+\phi_2}{2}\right) + y\sin\left(\frac{\phi_1+\phi_2}{2}\right) = a\cos\left(\phi_1 - \frac{\phi_1+\phi_2}{2}\right)$$

$$\Rightarrow x\cos\frac{1}{2}(\phi_1+\phi_2) + y\sin\frac{1}{2}(\phi_1+\phi_2) = a\cos\frac{1}{2}(\phi_1+\phi_2). \quad \textbf{Ans.}$$

Example 80:

Find the equation to the straight line passing through the pairs of points (a, b) and (a + b, a − b).

Solution:

Points are (a, b) and (a + b, a − b). Hence the equation is

$$y - b = \frac{(a-b)-(b)}{(a+b)-(a)}(x - a)$$

$$\Rightarrow \quad y - b = \frac{a-2b}{b}(x - a)$$

$$\Rightarrow \quad (a - 2b)x - by + b^2 + 2ab - a^2 = 0$$

Example 80(a):

Find the equation to the straight line passing through the pairs of points

$$\left(at_1^2, 2at_1\right) \text{ and } \left(at_2^2, 2at_2\right)$$

Solution:

The points are $\left(at_1^2, 2at_1\right)$ and $\left(at_2^2, 2at_2\right)$. Hence the equation is

$$y - 2at_1 = \frac{2at_2 - 2at_1}{at_2^2, - at_1^2}\left(x - at_1^2\right)$$

$$\Rightarrow \quad y - 2at_1 = \frac{2a(t_2 - t_1)\left(x - at_1^2\right)}{a(t_2 - t_1)(t_2 + t_1)}$$

$$\Rightarrow \quad y - 2at_1 = \frac{2a\left(x - at_1^2\right)}{a(t_2 + t_1)}$$

$$\Rightarrow \quad y(t_2 + t_1) - 2x = 2at_1t_2.$$

Example 80(b):

Find the equation to the straight line passing through the following pairs of points $\left(at_1, \frac{a}{t_1}\right)$ and $\left(at_2, \frac{a}{t_2}\right)$.

Solution:

The points are $\left(at_1, \frac{a}{t_1}\right)$ and $\left(at_2, \frac{a}{t_2}\right)$. Hence the equation is

$$y - \frac{a}{t_1} = \frac{(a/t_2) + (a/t_1)}{a/t_2 - a/t_1} \times (x - a/t_1)$$

$$\Rightarrow \quad \frac{yt_1 - a}{t_1} = \frac{a(t_1 - t_2)}{a(t_2 - t_1)\, t_1 t_2}(x - at_1)$$

$$\Rightarrow \quad yt_1t_2 - at_2 = -(x - at_1) \text{ or } yt_1t_2 + x = a\,(t_1 + t_2).$$ **Ans.**

Example 80(c):

Find the equations to the diagonals of the rectangle the equations of whose sides are x = a, x = a', y = b, and y = b'.

Solution:

Let x = a cuts the lines y = b and y = b' at A and D respectively. The co-ordinates of A will be (a, b) and of D will be (a, b). Similarly if x = a' cuts y = b and y = b' at B and C respectively, their co-ordinates will be (a', b) and (a', b'). Therefore the equation of the diagonal AC will be given as

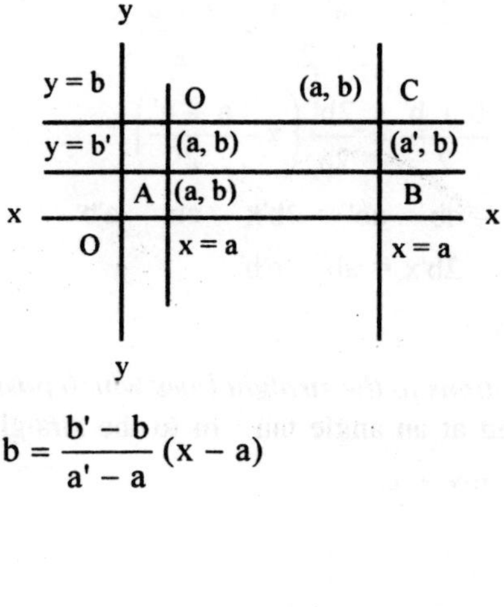

y
y = b
O
(a, b)
C
y = b'
(a, b)
(a', b)
A
(a, b)
B
x
x
O
x = a
x = a
y

$$y - b = \frac{b' - b}{a' - a}(x - a)$$

$\Rightarrow$ y (a' – a) – b (a' – a) = x (b' – b) – a (b' – b)

$\Rightarrow$ y (a' – a) – x (b' – b) = a'b – ab' **Ans.**

and the equation of the diagonal BD will be given as

$$y - b = \frac{b' - b}{a - a'}(x - a')$$

$\Rightarrow$ y (a – a') – b (a – a') = x (b' – b) – a' (b' – b)

$\Rightarrow$ x (b' – b) – y (a – a') = a'b' + a'b – ba + ba'

$\Rightarrow$ x (b' – b) + y (a' – a) = a'b' – ab. **Ans.**

Example 81(a):

Find the equation to the straight line which bisects the distance between the points (a, b) and (a', b') and also bisects the distance between the points (–a, b) and (a', –b').

Solution:

The co-ordinates of the mid-point (say P) of the line joining (a, b) are $\left(\frac{a + a'}{2}, \frac{b + b'}{2}\right)$.

The co-ordinates of the mid-point (say Q) of the line joining (–a, b) and (a', –b') are $\left(\frac{a' - a}{2}, \frac{b - b'}{2}\right)$.

The equation of the line joining P and Q is given by

$$y - \frac{b + b'}{2} = \frac{\frac{b - b'}{2} - \frac{b + b'}{2}}{\frac{a' - a}{2} - \frac{a + a'}{2}}\left(x - \frac{a + a'}{2}\right)$$

$$\Rightarrow \quad y - \frac{b + b'}{2} = \frac{2b'}{2a}\left(x - \frac{a + a'}{2}\right)$$

$\Rightarrow$ 2ay – ab – ab' = 2b'x – ab' – a'b'

$\Rightarrow$ 2ay – 2b'x = ab – a'b'. **Ans.**

Example 81(b):

Find the equations to the straight lines which pass through the point (h, k) and are inclined at an angle $\tan^{-1}$ m to the straight line

y = mx + c.

Solution:

The equations of the lines passing through (h, k) inclined at angle of $\tan^{-1}$ m to the straight line y = mx + c are

$$\text{(a)}\quad y-k = \frac{m+m}{1+m.m}(x-h)$$

[∵ m of given line is m and tan α = m]

$$\Rightarrow \qquad y - k = 0 \text{ or } y = k. \qquad \textbf{Ans.}$$

$$\text{(b)}\quad y-k = \frac{m+m}{1-m.m}(x-h)$$

$$\Rightarrow \qquad (1 - m^2)(y - k) = 2m(x - h). \qquad \textbf{Ans.}$$

Example 81(c):

Find the angle between the two straight lines 3x = 4y + 7 and 5y = 12x + 6 and also the equations to the two straight lines which pass through the point (4, 5) and make equal angles with the two given lines.

Solution:

The given lines are 3x = 4y = 7 ...(1)

and 5y = 12x + 6. ...(2)

Slope of the first line = $m_1 = \frac{3}{4}$ slope of second line $m^2 = \frac{12}{5}$

If angle between (1) and (2) be θ, then

$$\tan\theta = \pm\frac{m_1 - m_2}{1+m_1m_2} = \pm\frac{\frac{3}{4}-\frac{12}{5}}{1+\frac{3}{4}\frac{12}{5}} = \pm\frac{-33}{56}$$

Hence the acute angle between (1) and (2) is $\tan^{-1}\left(\frac{-33}{56}\right)$. **Ans.**

Again, any line passing through (4, 5) may be given by

$$y - 5 = m(x - 4). \qquad ...(3)$$

Let φ be the angle that the line (3) makes with (1) and (2) separately. Then

$$\tan\phi = \pm\frac{m-\frac{3}{4}}{1+m.\frac{3}{4}} = \pm\frac{4m-3}{4+3m} \qquad \text{[taking line (1) and (3)]}$$

Again $\tan\phi = \pm \dfrac{m - \frac{12}{5}}{1 + m.\frac{12}{5}} = \pm \dfrac{5m - 12}{5 + 12m}$ [taking line (2) and (3)].

Hence $\dfrac{4m - 3}{4 + 3m} = -\dfrac{5m - 12}{5 + 12m}$

[Taking one +ve and the other –ve sign, for ϕ will be +ve for one line and –ve for the other line]

$20m = 15 + 48m^2 - 36m = -20m - 15m^2 + 48 + 36m$

$63m^2 - 32m - 63 = 0$ or $(7m - 9)(9m + 7) = 0.$

So $m = \dfrac{9}{7}$ or $-\dfrac{7}{9}$

Substituting in (3), we get $y - 5 = \dfrac{9}{7}(x - 4)$ or $9x - 7y = 1$ and

$$y - 5 = -\frac{9}{7}(x - 4) \text{ or } 7x + 9y = 73$$ **Ans.**

Example 82:

Find the centre and radius of the circle which is inscribed in the triangle formed by thee straight lines whose equations are $2x + 4y + 3 = 0$, $4x + 3y + 3 = 0$, and $x + 1 = 0$.

Solution:

The equations of the sides of the Δ are given to be

$$2x + 4y + 3 = 0 \quad ...(1)$$

$$4x + 3y + 3 = 0 \quad ...(2)$$

and $$x + 1 = 0. \quad ...(3)$$

Let these equations represent BC, CA and AB respectively.

Solving (2) and (3), the co-ordinates of A are $\left(-1, \dfrac{1}{3}\right)$.

Solving (3) and (1), the co-ordinates of B are $\left(-1, \dfrac{1}{4}\right)$.

Solving (1) and (2), the co-ordinates of C are $\left(-\dfrac{3}{10}, -\dfrac{3}{5}\right)$.

Distance between B and C

i.e. $$a = \sqrt{\left\{\left(-\frac{3}{10}+1\right)^2 + \left(-\frac{3}{5}+\frac{1}{4}\right)^2\right\}} = \frac{7\sqrt{5}}{20}.$$

Distance between C and A

i.e. $$b = \sqrt{\left\{\left(-\frac{3}{10}+1\right)^2 + \left(-\frac{3}{5}-\frac{1}{3}\right)^2\right\}} = \frac{7}{6}.$$

Distance between A and B

i.e. $$c = \sqrt{\left\{(-1+1)^2 + \left(\frac{1}{3}+\frac{1}{4}\right)^2\right\}} = \frac{7}{12}.$$

Co-ordinates of the in-centre will be

$$\left[\left\{\frac{\left(\frac{7\sqrt{5}}{20}\times -1\right)+\left(\frac{7}{6}\times -1\right)+\left(\frac{7}{12}\times -\frac{3}{10}\right)}{\frac{7\sqrt{5}}{20}+\frac{7}{6}+\frac{7}{12}}\right\}\right.$$

$$\left.\left\{\frac{\left(\frac{7\sqrt{5}}{20}\times \frac{1}{3}\right)+\left(\frac{7}{6}\times -\frac{1}{4}\right)+\left(\frac{7}{12}\times -\frac{3}{3}\right)}{\frac{7\sqrt{5}}{20}+\frac{7}{6}+\frac{7}{12}}\right\}\right]$$

$$= \left[\frac{-85-7\sqrt{5}}{120}, \frac{21\sqrt{5}-65}{120}\right]$$ **Ans.**

Radius = distance from the centre of the line x + 1 = 0.

$$\left(1+\frac{-85-7\sqrt{5}}{12}\right) = \frac{35-7\sqrt{5}}{120} \text{ unit.}$$ **Ans.**

Example 83:

Prove that the co-ordinates of the centre of the circle inscribed in the triangle where angular points are (1, 2), (2, 3) and (3, 1) are

$(8+\sqrt{10})/6$ and $(16-\sqrt{10})/6$.

Solution:

The angular points are (1, 2), (2, 3) and (3, 1). Let these points be A, B and C respectively, then

The length BC = a = $\sqrt{[(3-2)^2 + (1-3)^2]} = \sqrt{5}$,

The length CA = b = $\sqrt{[(3-1)^2 + (1-2)^2]} = \sqrt{5}$,

The length AB = c = $\sqrt{[(2-1)^2 + (3-2)^2]} = \sqrt{2}$,

$\therefore$ The co-ordinates of in centre are

$$\left[\frac{\sqrt{5}\times 1+\sqrt{5}\times 2+\sqrt{2}\times 3}{\sqrt{5}+\sqrt{5}+\sqrt{2}}, \frac{\sqrt{5}\times 2+\sqrt{5}\times 3+\sqrt{2}\times 1}{\sqrt{5}+\sqrt{5}+\sqrt{2}}\right]$$

$$=\left[\frac{3\sqrt{5}+3\sqrt{2}}{2\sqrt{5}+\sqrt{2}}, \frac{5\sqrt{5}+\sqrt{2}}{2\sqrt{5}+\sqrt{2}}\right] \text{ or } \left(\frac{8+\sqrt{(10)}}{6}, \frac{6-\sqrt{(10)}}{6}\right)$$ **Ans.**

(on rationalization)

Co-ordinates of the centre of ex-circle opposite to A

$$=\left(\frac{-\sqrt{5}+2\sqrt{5}+3\sqrt{2}}{-\sqrt{5}+\sqrt{5}+\sqrt{2}}, \frac{-2\sqrt{5}+3\sqrt{5}+\sqrt{2}}{-\sqrt{5}+\sqrt{5}+\sqrt{2}}\right)$$

or $$\left(\frac{6+\sqrt{(10)}}{2}, \frac{2+\sqrt{(10)}}{6}\right).$$ **Ans.**

Similarly the co-ordinates of the centre of the circle opposite to vertex B are

$$\left(\frac{1\sqrt{5}-2\sqrt{5}+3\sqrt{2}}{\sqrt{5}-\sqrt{5}+\sqrt{2}}, \frac{2\sqrt{5}-3\sqrt{5}+\sqrt{2}}{\sqrt{5}-\sqrt{5}+\sqrt{2}}\right)$$

or $$\left(\frac{6-\sqrt{(10)}}{2}, \frac{2-\sqrt{(10)}}{2}\right).$$ **Ans.**

And the co-ordinates of the centre of the circle, opposite to vertex C are

$$\left(\frac{1\sqrt{5}+2\sqrt{5}-3\sqrt{2}}{\sqrt{5}+\sqrt{5}-\sqrt{2}}, \frac{2\sqrt{5}-3\sqrt{5}+\sqrt{2}}{\sqrt{5}+\sqrt{5}-\sqrt{2}}\right)$$

or $$\left(\frac{8-\sqrt{(10)}}{6}, \frac{16+\sqrt{(10)}}{6}\right).$$ **Ans.**

Example 84(a):

Find the co-ordinates of the centres, and the radii, of the four circles which touch the sides of the triangle the co-ordinates of whose angular points are the points (6, 0) (0, 6) and (7, 7).

Solution:

The co-ordinates of the points are given as (6, 0), (0, 6) and (7, 7).

Let three points A, B and C respectively; then

The length BC = a = $\sqrt{[(7-0)^2 + (7-6)^2]} = 5\sqrt{2}$,

The length CA = b = $\sqrt{[(7-6)^2 + (7-0)^2]} = 5\sqrt{2}$,

The length AB = c = $\sqrt{[(6-0)^2 + (0-6)^2]} = 6\sqrt{2}$.

The co-ordinates of in-centre are

$$\left[\frac{(6\times 5\sqrt{2}) + (0\times 5\sqrt{2}) + (7\times 6\sqrt{2})}{5\sqrt{2} + 5\sqrt{2} + 6\sqrt{2}}, \frac{5\sqrt{2}\times 0 + 6\times 5\sqrt{2} + 7\times 6\sqrt{2}}{5\sqrt{2} + 5\sqrt{2} + 6\sqrt{2}}\right]$$

or $\left(\frac{9}{2}, \frac{9}{2}\right)$.

Equation of the line passing through (6, 0) and (0, 6)

$$\frac{x}{6} + \frac{y}{6} = 1 \text{ or } x + y - 6 = 0.$$

Radius of in circle $= \dfrac{\frac{9}{2} + \frac{9}{2} - 6}{\sqrt{(1+1)}} = \dfrac{3}{\sqrt{2}} = \dfrac{3\sqrt{2}}{2}$ **Ans.**

Centre of the circle opposite to vertex A

$$= \left(\frac{-6\times 5\sqrt{2} + 0 + 7\times 6\sqrt{2}}{-5\sqrt{2} + 5\sqrt{2} + 6\sqrt{2}}, \frac{-0 + 30\sqrt{2} + 42\sqrt{2}}{-5\sqrt{2} + 5\sqrt{2} + 6\sqrt{2}}\right)$$

= (2, 13). **Ans.**

Example 84(b):

Two fixed straight lines OX and OY are cut by a variable line in the point A and B respectively and P and Q are the feet of the perpendiculars drawn from A and B upon the lines OBY and OAX. Show that, if AB passes through a fixed point, then PQ will also pass through a fixed point.

Solution:

Let OX be the x-axis and OY be any line passing through origin; hence its equation may be given by

$$y = mx \qquad \ldots(1)$$

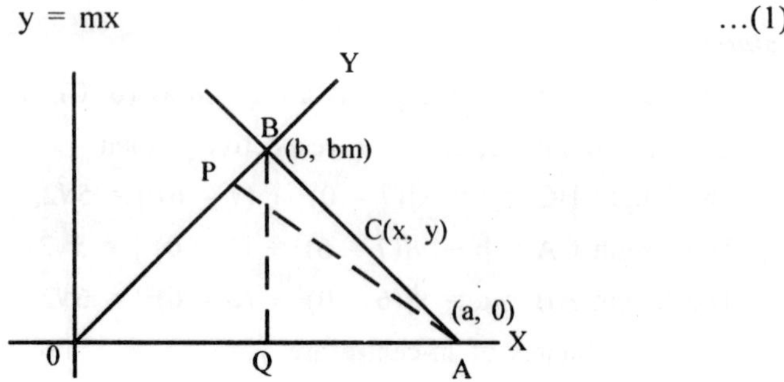

Therefore co-ordinates of A may be given by (a, 0) and B (b, bm). If the straight line AB always passes through a fixed point, say C, then slope of AC and BC are same. Hence

$$\frac{y_1 - 0}{x_1 - a} = \frac{y_1 - bm}{x_1 - b}$$

or $\qquad y_1x_1 - by_1 = y_1x_1 = bmx_1 - ay_1 + abm$

or $\qquad b(mx_1 - y_1) + ay_1 - abm = 0 \qquad \ldots(2)$

Equation of AP will be $x - my = a$. $\qquad \ldots(3)$

[as PA ⊥ OY and passes through (a, 0)]

Solving (1) and (3), we get the co-ordinates of $P = \left(\frac{a}{1+m^2}, \frac{am}{1+m^2}\right)$.

Again if BQ is perpendicular to OX, it will be parallel to y-axis and as it passes through (b, bm), its equation will be x = b and therefore the co-ordinates of Q will be (b, 0).

The equation of the line passing through

$$P = \left(\frac{a}{1+m^2}, \frac{am}{1+m^2}\right) \text{ and } Q = (b, 0)$$

is $\qquad y - 0 = \dfrac{\{am/1+m^2\} - 0}{\{a/1+m^2\} - b}(x - b)$

or $\qquad y\{a - b(1 + m^2)\} = am(x - b)$

or $\quad b(y + ym^2) + a(mx - y) - abm = 0.$ ...(4)

Subtracting (2) from (4), we get

$$b(y + ym^2 - mx_1 + y_1) + a(mx - y - y_1) = 0$$

or
$$\left(y + ym^2 - mx_1 + y_1\right) + \frac{a}{b}\left(mx - y - y_1\right) = 0$$

This equation is of the form $P + \lambda Q = 0$ (l being equal to a/b).

Hence it passes through a fixed point which is the point of intersection of $y + ym^2 - mx_1 + y_1 = 0$ and $mx - y - y_1 = 0$. **Proved.**

Note: This equation can also be done by taking OX and OY as axes of x and ω inclined at an angle w, points A and B being, (a, 0) and **(0, b)** respectively.

Example 85:

*A straight line is such that the algebraic sum of the perpendiculars **let** fall upon it from any number of fixed points is zero; show that it always **passes** through a fixed point.*

Solution:

Let the equation of the straight line be $ax + by + c = 0$ ...(1)

and the points be (x_1, y_1), (x_2, y_2), ...(x_n, y_n). Then by hypothesis,

$$\frac{ax_1 + by_1 + c}{\sqrt{(a^2 + b^2)}} + \frac{ax_2 + by_2 + c}{\sqrt{(a^2 + b^2)}} + \ldots + \frac{ax_n + by_n + c}{\sqrt{(a^2 + b^2)}} = 0$$

or $\quad a\sum_{n=1}^{n} x_n + b\sum_{y=1}^{n} y_n + nc = 0.$ ...(2)

Multiplying equation no. (1) by n and subtracting (2) from it, we get

$$a[nx - \Sigma x_n] + b[ny - \Sigma y_n] = 0$$

or
$$[nx - \Sigma x_n] + \tfrac{b}{a}[ny - \Sigma y_n] = 0$$

This equation is of the form $P + \lambda Q = 0$ as $l = \frac{b}{a}$ = constant.

Hence the line passes through a fixed point which is the point of intersection of $nx - \Sigma x_n = 0$ and $ny - \Sigma y_n = 0$.

or
$$\left[\frac{\Sigma x_n}{n}, \frac{\Sigma y_n}{n}\right]$$
Proved.

Example 86:

If a straight line move so that the sum of the perpendiculars let fall on it from the two fixed points (3, 4) and (7, 2) is equal to three times the perpendicular on it from a third fixed point (1, 3), prove that there is another fixed point through which this line always passes and find its co-ordinates.

Solution:

Let the equation of the straight line be

$$ax + by + c = 0. \quad ...(1)$$

As sum of the lengths of perpendiculars from (3, 4) and (7, 2) is equal to 3 times the length of the perpendicular from (1, 3), we get

$$\frac{3a + 4b + c}{\sqrt{(a^2 + b^2)}} + \frac{7a + 2b + c}{\sqrt{(a^2 + b^2)}} = 3.\frac{1.a + 3b + c}{\sqrt{(a^2 + b^2)}}$$

or $$7a - 3b - c = 0 \quad ...(2)$$

Adding (1) and (2), we get

$$a(x + 7) + b(y - 3) = 0$$

or $$(x + 7) + \frac{b}{a}(y - 3) = 0$$

As this equation is of the form $P + \lambda Q = 0$, it passes through the fixed point which is the point of intersection of $x + 7 = 0$ and $y - 3 = 0$.

The required point is (–7, 3). **Ans.**

Example 87:

If the equal sides AB and AC of an isosceles triangle be produced to E and F so that BE.CE = AB^2, show that the line EF will always pass through a fixed point.

Solution:

Let us suppose the lengths of equal sides be a and taking AB and AC as axes of x and y respectively, we get the co-ordinates of B as (a, 0) and of C as (0, a). Again let AE = h and AF = k. Then the equation of EF will be

$$\frac{x}{h} + \frac{y}{k} = 1 \quad ...(1)$$

Given condition is BE.CF = AB^2

or $$(h - a)(k - a) = a^2$$

or $\quad hk - ah - ak + a^2 = a^2$

or $\quad ah + ak = hk$ or $\dfrac{a}{k} + \dfrac{a}{h} = 1$ ...(2)

Comparing (1) and (2), we find that (a, a) always satisfies equation No. (1). Hence (1) always passes through (a, a) which is a fixed point.

Example 88(a):

Find the centre and radius of the circle which is inscribed in the triangle formed by the straight lines whose equations are y = 0, 12x – 5y = 0, and 3x + 4y – 7 = 0.

Solution:

The lines are

$$y = 0, \quad ...(1)$$

$$12x - 5y = 0, \quad ...(2)$$

and $$3x + 4y - 7 = 0. \quad ...(3)$$

Solving (2) and (3), the co-ordinates of A are $= \left(\dfrac{5}{9}, \dfrac{4}{3}\right)$.

Solving (3) and (1), the co-ordinates of B are $= \left(\dfrac{7}{9}, 0\right)$.

Solving (1) and (2), the co-ordinates of C are = (0, 0)

$$\text{Distance BC} = a = \sqrt{\left[\left(\frac{7}{3} - 0\right)^2 - (0)^2\right]} = \frac{7}{3}.$$

$$\text{Distance CA} = b = \sqrt{\left[\left(\frac{5}{9} - 0\right)^2 + \left(\frac{4}{3} - 0\right)^2\right]} = \frac{12}{9}.$$

$$\text{Distance AB} = c = \sqrt{\left[\left(\frac{5}{9} - \frac{7}{3}\right)^2 + \left(\frac{4}{3} - 0\right)^2\right]} = \frac{20}{9}.$$

Co-ordinates of in centre will be

$$\left[\frac{\frac{7}{3} \times \frac{5}{9} + \frac{13}{9} \times \frac{7}{3} + \frac{20}{9} \times 0}{\frac{7}{3} + \frac{13}{9} + \frac{20}{9}}, \frac{\frac{7}{3} \times \frac{4}{3} + \frac{13}{9} \times 0 + \frac{20}{9} \times 0}{\frac{7}{3} + \frac{13}{9} + \frac{20}{9}}\right] = \left[\frac{7}{9}, \frac{14}{27}\right]$$

Radius = perpendicular from centre on x-axis,

i.e. $y = 0$ will be $\frac{14}{27}$. **Ans.**

Example 88(b):

Find the equations to the sides and diagonals of a regular hexagon, two of its sides, which meet in a corner, being the axes of co-ordinates.

Solution:

Let AB and AF represent axes of x and y respectively. Then the equation of AB is clearly

$$y = 0. \qquad ...(1)$$

If the side of the hexagon be a, clearly by the Fig., the intercepts made by BC on axes are a and –a.

Hence its equation will be

$$\frac{x}{a} - = \frac{y}{a} = 1 \text{ or } x - y = a \qquad ...(2)$$

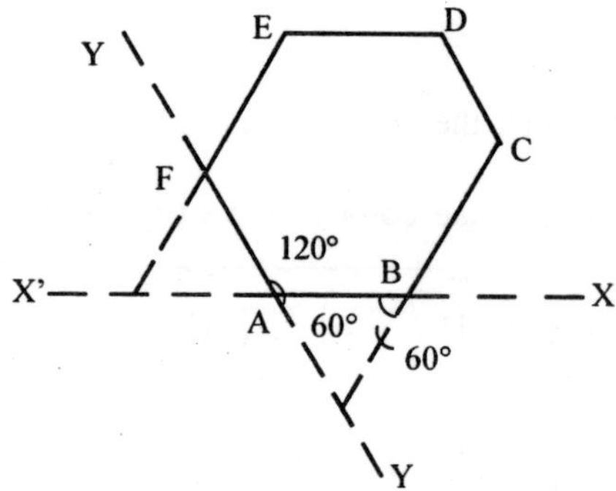

CD is parallel to y-axis and if produced cuts x-axis at a distance of 2a; hence its equation will be

$$x = 2a. \qquad ...(3)$$

Similarly DE is parallel to x-axis and if produced meets y-axis at a distance of 2a, hence its equation will be

$$y = 2a. \qquad ...(4)$$

EF cuts intercepts of –a and a from axes; hence its equation will be

$$\frac{x}{-a} - = \frac{y}{a} = 1 \text{ or } x - y + a = 0. \qquad ...(5)$$

FA is clearly y-axis, hence its equation is $x = 0$. ...(6)

Diagonal AD bisects the angle between the axes; hence its equation will be $x - y = 0$. ...(7)

Diagonal BE is parallel to y-axis at a distance of a units, hence its equation is $x = a$. ...(8)

Diagonal CF is parallel to x-axis at a distance of a units, so its equation is $y = a$. ...(9)

Hence the equations (1) to (9) represent the required equations.

Example 88(c):

From a given point (h, k) perpendiculars are drawn to the axes, whose inclination is w, and their feet are joined. Prove that the length of the perpendicular drawn from (h, k) upon this line is $\dfrac{hk \sin^2 \omega}{\sqrt{h^2 + k^2 + 2hk \cos \omega}}$, *and that its equation is* $hx - ky = h^2 - k^2$.

Solution:

In the Fig. of Q. No. 163, let PN be the perpendicular from P on QK, then

$$\text{Area of } \Delta PQK = \frac{1}{2} QK.PN, \qquad ...(1)$$

and again taking PK base and Q vertex,

$$\text{Area of } \Delta PQK = \frac{1}{2} PK.PQ \sin QPK. \qquad ...(2)$$

As $\angle QPK = (180° - \omega)$, by (1) and (2), we get

$$\frac{1}{2} QK.PN = \frac{1}{2}.PK.PQ, \sin (180° - \omega)$$

or $[\sin w \sqrt{(h^2 + k^2 - 2hk \cos \omega)}]\, PN = k \sin \omega . h \sin \omega \sin \omega.$

$$\therefore \quad PN = \frac{hk \sin^2 \omega}{\sqrt{(h^2 + k^2 - 2hk \cos \omega)}}.$$ **Proved.**

Again equation of QK is $\dfrac{x}{OK} + \dfrac{y}{OQ} = 1$,

$$\frac{x}{h + k \cos \omega} + \frac{y}{k + h \cos \omega} = 1. \qquad ...(3)$$

Equation of any line through (h, k) will be

$$y - k = m (x - h). \qquad ...(4)$$

As (3) and (4) are perpendicular to each other

$$1 + \left(m - \frac{k + h\cos\omega}{h + k\cos\omega}\right)\cos\omega - \frac{m(k + h\cos\omega)}{hk\cos\omega} = 0$$

Solving we get m = h/k.

Substituting in (4), we get $y - k = \frac{h}{k}(x - h)$

or $hx - ky = h^2$ (on simplification). **Proved.**

Example 89(a):

Find the length of the perpendicular drawn from the point (4, –3) upon the straight line 6x + 3y – 10 = 0, the angle between the axes being 60°.

Solution:

The point is given as (4, –3) and the line is 6x + 3y – 10 = 0.

Substituting the values of a, b, c, x, y and ω, we get

$$P = \frac{6.4 + 3.(-3) + (-10)}{\sqrt{(6^2 + 3^2 - 2.6.3\cos 60^\circ)}}\sin 60^\circ = \frac{5}{6}.$$ **Ans.**

Example 89(b):

Find the equation to, and the length of, the perpendicular drawn from the point (1, 1) upon the straight line 3x + 4y + 5 = 0, the angle between the axesbeing 120°.

Solution:

The length of perpendicular from the point (1, 1) upon the given line 3x + 4y + 5 = 0 will be

$$P = \frac{3.1 + 4.1 + 5}{\sqrt{(3^2 + 4^2 - 2.3.4\cos 120^\circ)}}\sin 120^\circ = \frac{6}{37}\sqrt{(111)}.$$ **Ans.**

Again equation of any line through (1, 1) will be

$$y - 1 = m(x - 1). \qquad ...(1)$$

As the required line (1) is perpendicular to the line 3x + 4y + 5 = 0, we have

$$1 + \left(m - \frac{3}{4}\right)\cos 120 - \frac{3}{4}m = 0 \text{ or } m = \frac{11}{10}.$$

Example 90:

Find the centre and radius of the circle which is inscribed in the triangle formed by the straight lines whose equations are $3x + 4y + 2 = 0$, $3x - 4y + 12 = 0$, and $4x - 3y = 0$.

Solution:

The equations of the sides of the triangle are given

$$3x + 4y + 2 = 0 \qquad ...(1)$$

$$3x - 4y + 12 = 0 \qquad ...(2)$$

and $$4x - 3y = 0 \qquad ...(3)$$

Let these equations represent BC, CA and AB respectively.

Solving (2) and (3), the co-ordinates of A are $\left(\frac{36}{7}, \frac{48}{7}\right)$.

Solving (3) and (1), the co-ordinates of B are $\left(-\frac{6}{25}, -\frac{8}{25}\right)$.

Solving (1) and (2), the co-ordinates of C are $\left(-\frac{7}{3}, \frac{5}{4}\right)$.

Distance between B and C

i.e. $$a = \sqrt{\left\{\left(-\frac{6}{25} + \frac{7}{3}\right)^2 + \left(-\frac{8}{25} - \frac{5}{4}\right)^2\right\}} = \frac{157}{60}.$$

Distance between C and A

i.e. $$b = \sqrt{\left\{\left(\frac{36}{7} + \frac{7}{3}\right)^2 + \left(\frac{48}{7} - \frac{5}{4}\right)^2\right\}} = \frac{785}{84}.$$

Distance between A and B

i.e. $$c = \sqrt{\left\{\left(\frac{36}{7} + \frac{6}{25}\right)^2 + \left(\frac{48}{7} + \frac{8}{25}\right)^2\right\}} = \frac{314}{35}.$$

Co-ordinates of the in-centre will be

$$\left[\left\{\frac{\frac{157}{60} \times \frac{36}{7} + \frac{785}{24} \times \frac{-6}{25} + \frac{314}{35} \times \frac{-7}{3}}{\frac{157}{60} + \frac{785}{84} + \frac{314}{35}}\right\}\right.$$

$$\left[\left\{\frac{\frac{157}{60}\times\frac{48}{7}+\frac{785}{84}\times\frac{-8}{25}+\frac{314}{35}\times\frac{5}{4}}{\frac{157}{60}+\frac{785}{84}+\frac{314}{35}}\right\}\right]=\left(-\frac{13}{28},\frac{5}{4}\right)$$ **Ans.**

Radius = Perpendicular on AB from centre

$$=\frac{-4\times\frac{13}{28}-3\times\frac{5}{4}}{\sqrt{(9+16)}}=\frac{157}{140}\text{ unit.}$$ **Ans.**

Example 91:

From each corner of a parallelogram of perpendicular is drawn upon the diagonal which does not pass through that corner and these are produced to form another parallelogram; show that its diagonals are perpendicular to the sides of the first parallelogram and that they both have the same centre.

Solution:

Let ABCD be the parallelogram. Taking diagonals AC and BD as the axes & of lengths 2a and 2b respectively, co-ordinates of A, B, C and D will be (–a, 0), (0, –b), (a, 0) respectively.

Equations of DK and BL will be respectively as

$$x \sec \omega + y = b \qquad ...(1)$$

and $$x \sec \omega + y = -b \qquad ...(2)$$

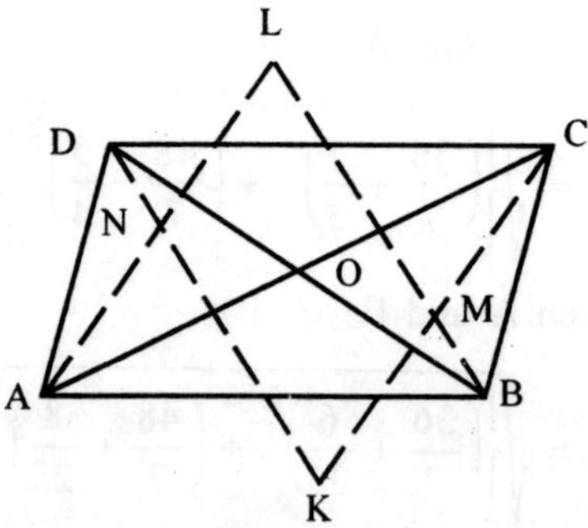

Similarly equations of CK and AL will be respectively as

$$x + y \sec \omega = a \qquad ...(3)$$

and $$x + y \sec \omega = -a. \qquad ...(4)$$

Multiplying (1) by a and (4) by b and adding, we get the equation

$$x\,(a \sec \omega + b) + y\,(a + b \sec \omega) = 0. \qquad \ldots(5)$$

This represents any line passing through N, the point of intersection of AL and DK.

Similarly multiplying (2) by a and (3) by b and adding, we get the equation $x\,(a \cos \omega + b) + y\,(a + b \sec \omega) = 0$, which represents the line passing through M, the point of intersection of CK and BL. As this is same as (5), equation no. (5) represents the equation of NM.

Similarly the equation of LK will be

$$x\,(a \sec \omega - b) + y\,(a - b \sec \omega) = 0 \qquad \ldots(6)$$

Solving (5) and (6), we get

$$x = 0,\ y = 0 \text{ (as } a + b \neq 0,\ 1 + \sec \omega \neq 0).$$

Therefore diagonals of KMLN meet in origin, which is same as the meeting point of the diagonals of ABCD. **Proved.**

Again equation of DC is clearly $\dfrac{x}{a} + \dfrac{y}{b} = 1.$ ...(7)

($\because$ it cuts the intercepts of a and b from the axis)

The lines CD and KL given by equations (7) and (6) are clearly perpendicular to each other as they satisfy the required condition.

Example 92:

The axes being inclined at an angle of 30°, find the equation to the straight line which passes through the point (–2, 3) and is perpendicular to the straight line y + 3x = 6.

Solution:

Any line through the point (–2, 3) is

$$y - 3 = m\,(x + 2) \qquad \ldots(1)$$

As the line is perpendicular to they + 3x = 6, we have

$$1 + (m - 3) \cos 30° - 3m = 0 \text{ or } m = \frac{3\sqrt{3} - 2}{\sqrt{3} - 6}.$$

Substituting in (1), the required line is

$$y - 3 = \frac{3\sqrt{3} - 2}{\sqrt{3} - 6}\,(x + 2).$$ **Ans.**

Example 93(a):

The axes being inclined a tan angle of 120°, find the tangent of the angle between the two straight lines 8x + 7y = 1 and 28x – 73y = 101.

Solution:

Lines are given as

$$8x + 7y = 1 \quad \ldots(1)$$

and $$28x - 73y = 101. \quad \ldots(2)$$

$\therefore$ $$m_1 = -\frac{8}{7},\ m_2 = \frac{28}{73}$$

and $$\omega = 120° \text{ (given)}$$

$$\therefore \quad \tan\theta = \frac{\left(-\frac{8}{7} - \frac{28}{73}\right)\sin 120°}{1 + \left(-\frac{8}{7} + \frac{28}{73}\right)\cos 120° + \left(-\frac{8}{7}\right)\left(\frac{28}{73}\right)} = \frac{30\sqrt{3}}{37}.$$

$$\therefore \quad \theta = \tan^{-1}\left(\frac{30\sqrt{3}}{37}\right).$$

Example 93(b):

Prove that the straight lines y + x = c and y = x + d are at right angles, whatever be the angle between the axes.

Solution:

Two lines are perpendicular to each other, if

$$1 + (m_1 + m_2)\cos\omega + m_1 m_2 = 0 \ldots .(1)$$

According to the question, $m_1 = -1$, $m_2 = 1$. Substituting in (1), we get $1 + (-1 + 1)\cos\omega + (-1)(1) = 0$ for all values of ω.

Hence lines are perpendicular for all values of ω. **Proved.**

Example 93(c):

Prove that the equation to the straight line which passes through the point (h, k) and is perpendicular to the axis of x is

$$x + y\cos\omega = h + k\cos\omega.$$

Solution:

Equation of any straight line passing through (h, k) may be given by

$$y - k = m\,(x - h). \qquad \ldots(1)$$

If it is perpendicular to x-axis, then

$$1 + m\cos\omega = 0 \text{ or } m = -\frac{1}{\cos\omega}.$$

$$y\cos\omega - k\cos\omega = -x + h$$

$$x + y\cos\omega = h + k\cos\omega.$$ **Proved.**

Example 94(a):

If $y = x\tan\frac{11\pi}{24}$ *and* $y = x\tan\frac{19\pi}{24}$ *represent two straight lines at right angles prove that the angle between the axes is* $\pi/4$.

Solution:

The equation are given as $y = x\tan\frac{11\pi}{24}$. ...(1)

and $y = x\tan\frac{19\pi}{24}$. ...(2)

If two lines are perpendicular to each other and the axes be inclined at an angle of ω, the condition is $1 + (m + m')\cos\omega + mm' = 0$.

(Article 93 cor. 2)

Applying the condition

$$1 + \left(\tan\frac{11\pi}{24} + \tan\frac{19\pi}{24}\right)\cos\omega + \tan\frac{11\pi}{24} \text{ or } \frac{19\pi}{24} = 0$$

or $$\left[\frac{\sin\frac{11\pi}{24}\cos\frac{19\pi}{24} + \sin\frac{19\pi}{24}\cos\frac{11\pi}{24}}{\cos\frac{11\pi}{4}\,.\,\cos\frac{19\pi}{4}}\right]\cos\omega$$

$$= -\left[\frac{\cos\frac{11\pi}{24}\cos\frac{19\pi}{4} + \sin\frac{19\pi}{4}\,.\,\sin\frac{11\pi}{4}}{\cos\frac{11\pi}{4}\,.\,\cos\frac{19\pi}{4}}\right]$$

or $$\sin\left(\frac{11\pi}{24} + \frac{19\pi}{24}\right)\cos\omega = -\cos\left(\frac{19\pi}{24} - \frac{11\pi}{24}\right)$$

or $$\sin(5\pi/4)\cos\omega = -\cos(\pi/3)$$

or $$\cos\omega = -\frac{\cos(\pi/3)}{\sin(5\pi/4)} = -\frac{(1/2)}{(-1/\sqrt{2})} = \frac{1}{\sqrt{2}}$$

or $$\omega = \pi/4.$$ **Proved.**

Example 94(b):

With oblique co-ordinates find the tangent of the angle between the straight lines y = mx + c and my + x = d.

Solution:

Let w be the angle between the axes. The lines are given as

$$y = mx + c \qquad \text{...(1)}$$

and $$my + x = d \text{ or } y = -\frac{1}{m}x + \frac{d}{m}. \qquad \text{...(2)}$$

If θ be the angle between the lines, then

$$\tan\theta = \frac{\left[m - \left(-\frac{1}{m}\right)\right]\sin\omega}{1 + \left[m + \left(-\frac{1}{m}\right)\right]\cos\omega + m\left(-\frac{1}{m}\right)}$$

$$= \frac{\frac{m^2+1}{m}\sin\omega}{1 + \left[\frac{m^2+1}{m}\right]\cos\omega - 1} = \frac{m^2+1}{m^2-1}\tan\omega$$

$$\therefore \quad \tan\theta = \frac{m^2+1}{m^2-1}\tan\omega.$$ **Ans.**

Note: In answer they have given the value of θ while according to the question the value of tan θ is required.

The Straight Line, Polar Equations Oblique Co-Ordinates

Formulae:

In oblique co-ordinates the equation y = mx + c represents a straight line which is inclined at an angle

$$\tan^{-1}\left(\frac{m \sin \omega}{1 + m \cos \omega}\right)$$

to the x-axis, where ω is the angle between the axes.

Example 95:

The axes being inclined at an angle of 60°, find the inclination to the axis of x of the straight lines whose equations are y = 2x + 5 and 2y = (√3 – 1) x + 7.

Solution:

The angle between the axes is 60°. The line is given as y = 2x + 5.

(i) ∴ m = 2 and ω = 60°. If θ be the inclination of the line to x-axis,

$$\tan \theta = \frac{m \sin \omega}{1 + m \cos \omega} = \frac{2 \sin 60°}{1 + 2 \cos 60°} = \frac{\sqrt{3}}{2}.$$

$$\therefore \qquad \theta = \tan^{-1}\left(\frac{\sqrt{3}}{2}\right) \quad \textbf{Ans.}$$

(ii) The equation of the line is 2y = (√3 – 1) x + 7.

$\therefore \quad m = \dfrac{\sqrt{3} - 1}{2}$ and ω = 60° (given), hence if θ be the inclination of the line to x-axis,

$$\tan \theta = \frac{m \sin \omega}{1 + m \cos \omega} = \frac{[(\sqrt{3} - 1)/2] \sin 60°}{1 + [(\sqrt{3} - 1)/2] \cos 60°}$$

$$= \frac{\sqrt{3}\,(\sqrt{3} - 1)}{\sqrt{3}\,(\sqrt{3} + 1)} = \frac{\sqrt{3} - 1}{\sqrt{3} + 1}$$

is $\theta = 15°$. **Ans.**

Formula:

If θ be the angle between two lines $y = m_1x + c_1$ and $y = m_2x + c_2$, ω be the angle between the axes, then

$$\tan \theta = \frac{(m_1 - m_2) \sin \omega}{1 + (m_1 + m_2) \cos \omega + m_1 m_2}.$$

Example 96(a):

If the straight lines $y = m_1x + c_1 = m_2x + c_2$ *make equal angles with the axis of x and be not parallel to one another, prove that*

$$m_1 + m_2 + 2m_1m_2 \cos \omega = 0.$$

Solution:

If two lines make equal angles with x-axis and are not parallel, then only possibility is that angles are supplementary, i.e. if one line makes the angle θ with x-axis, the other will make = θ.

$$\therefore \qquad \tan^{-1} \frac{m_1 \sin \omega}{1 + m_1 \cos \omega} = -\tan^{-1} \frac{m_2 \sin \omega}{1 + m_2 \cos \omega}$$

$$\text{or} \qquad \frac{m_1 \sin \omega}{1 + m_1 \cos \omega} \cdot \frac{m_2 \sin \omega}{1 + m_2 \cos \omega} = 0$$

$$\text{or} \qquad m_1 + m_2 + 2m_1m_2 \cos \omega = 0 \quad (\because \sin \omega \neq 0).$$

Example 96(b):

A straight line AB, whose length is c, slides between two given oblique axes which meet at O; find the locus of the orthocentre of the triangle OAB.

Solution:

Taking O as origin, OA and OB as the axes, ∠ AOB = ω, OA = a and OB = b. Then

Equation of AL, the perpendicular from A on OB, will be

$$x \cos \omega + y = a \cos \omega. \qquad \ldots(1)$$

Similarly the equation of BM, the perpendicular from B on OA will be

$$x + y \cos \omega = b \cos \omega. \qquad \ldots(2)$$

Again as AB is given to be c, hence

$$c^2 = a^2 + b^2 - 2ab \cos \omega. \qquad \ldots(3)$$

Putting the values of a and b from (1) and (2) IN (3), we get

$$c^2 = (x + y \sec \omega)^2 + (y + x \sec \omega)^2$$

$$- 2 \cos \omega (x + y \sec \omega)(y + x \sec \omega)$$

$$\Rightarrow \qquad c^2 \cot^2 \omega = x^2 + y^2 + 2xy \cos \omega.$$

Example 96(c):

A variable straight line cuts off from n given concurrent straight lines intercepts the sum of the reciprocals of which is constant. Show that it always passes through a fixed point.

Solution:

Let the fixed point be the origin; the angles that the various fixed lines make with initial line be α, β, γ ... etc.

Let the variable line be $\frac{1}{r}$ = (a cos θ + b sin θ). Let it cut the given fixed lines at (r_1, θ), (r_2, θ), (r_3, θ), ... etc. They by hypothesis.

$$\frac{1}{r_1} + \frac{1}{r_2} + \frac{1}{r_3} + \dots \text{ constant.}$$

or $$(a \cos \alpha + b \sin \alpha) + (a \cos \beta + b \sin \beta) + \dots \text{ constant.}$$

or $$a\Sigma \cos \alpha + b\Sigma \sin \beta = \text{constant} = c \text{ (say).}$$

Hence the line passes through a fixed point which is

$$\left[\frac{\Sigma \cos \alpha}{c}, \frac{\Sigma \sin \alpha}{c}\right].$$ **Proved.**

Example 97:

Having given the bases and the sum of the area of a number of triangles which have a common vertex, show that the locus of this vertex is a straight line.

Solution:

Let the co-ordinates of the vertex be (h, k), and let the lengths of the bases bee l_1, l_2, l_3 ...etc. and their equations respectively as

$$x \cos \alpha + y \sin \alpha = p_1,$$

$$x \cos \beta + y \sin \beta = p_2, \text{ ...etc.}$$

Then the lengths of perpendiculars from (h, k) on bases are respectively (h cos α + k sin α – p_1), (h cos β + k cos β – p_2), ...etc.

Sum of the areas of the Δs is constant by hypothesis.

$$\therefore \quad \frac{1}{2} l_1(h \cos \alpha + k \sin \alpha - p_1) + \frac{1}{2} l_2 (h \cos \beta + k \cos \beta - p_2)$$

$$+ \text{ ...etc.} = \text{constant} = c/2 \text{ (say)}$$

or $$h (l_1 \cos \alpha + l_2 \cos \beta + \dots) + + (l_1 \sin \alpha + l_2 \sin \beta + \dots)$$

$$-(l_1 p_1 + l_2 p_2 + \dots) = c.$$

Generalising, we get, $x.\Sigma l_1 \cos \alpha + y.\Sigma l_1 \sin \alpha - \Sigma l_1 p_1 = c$.

It is an equation of a straight line. **Ans.**

Example 98:

Find the locus of a point at which two given portions of the same straight line subtend equal angles.

Solution:

Let OA and BC be the two portions of the straight line OABC. Taking O as origin, OC as axis of x, the co-ordinates of O, A, B and C may be taken respectively as (0, 0), (a, 0), (b, 0) and (c, 0). Say the point P is (h, k).

$$\text{Slope of OP} = m_1 = \frac{k}{h}.$$

$$\text{Slope of PA} = m_2 = \frac{k-0}{h-a}.$$

$$\text{Slope of PB} = m_3 = \frac{k-0}{h-b}.$$

$$\text{Slope of PC} = m_4 = \frac{k-0}{h-c}.$$

$$\text{Angles between OP and PA} = \tan^{-1}\left[\frac{\left(\frac{k}{h}-\frac{k}{h-a}\right)}{\left(1+\frac{k}{h}\cdot\frac{k}{h-a}\right)}\right].$$

$$\text{Angle between PB and PC} = \tan^{-1}\left[\frac{\left(\frac{k}{h-b}-\frac{k}{h-c}\right)}{\left(1+\frac{k}{h-b}\cdot\frac{k}{h-c}\right)}\right].$$

By hypothesis,
$$\frac{\left(\frac{k}{h}-\frac{k}{h-a}\right)}{\left(1+\frac{k}{h}\cdot\frac{k}{h-a}\right)} = \frac{\left(\frac{k}{h-b}-\frac{k}{h-c}\right)}{\left(1+\frac{k}{h-b}\cdot\frac{k}{h-c}\right)}$$

or
$$\frac{-ak}{h^2-ah+k^2} = \frac{kb-kc}{h^2-hc-hb+bc+k^2}.$$

Simplifying and generalising, we get

$$(a + b - c)(x^2 + y^2) - 2abx - abc = 0. \quad \textbf{Ans.}$$

Example 99:

Through a given point O a straight line is drawn to cut two given straight lines in R and S; find the locus of a point P on this variable straight line, which is such that (1) 2OP = OR + OS, and (2) OP² = OR.OS.

Solution:

Suppose the axes pass through O. Let the given fixed lines be NI and NM having the equations as

$$a_1x + b_1y + c_1 = 0 \quad \text{...(1)}$$

and

$$a_2x + b_2y + c_2 = 0 \quad \text{...(2)}$$

Transforming in polar equations, we get

$$\frac{1}{r} = \frac{a_1 \cos\theta + b_1 \sin\theta}{c_1} \text{ and } \frac{1}{r} = \frac{a_2 \cos\theta + b_2 \sin\theta}{c_2}$$

Taking OR and OS as r_1 and r_2 respectively and $\angle XOR = \theta$, we get

$$r_1 = \frac{c_1}{a_1 \cos\theta + b_1 \sin\theta} \text{ and } r_2 = \frac{c_2}{a_2 \cos\theta + b_2 \sin\theta}$$

(1) If OP = r, then by hypothesis 2OP = OR + OS or $2r = r_1 + r_2$.

or

$$2r = \left[\frac{c_1}{a_1 \cos\theta + b_1 \sin\theta} + \frac{c_2}{a_2 \cos\theta + b_2 \sin\theta}\right].$$

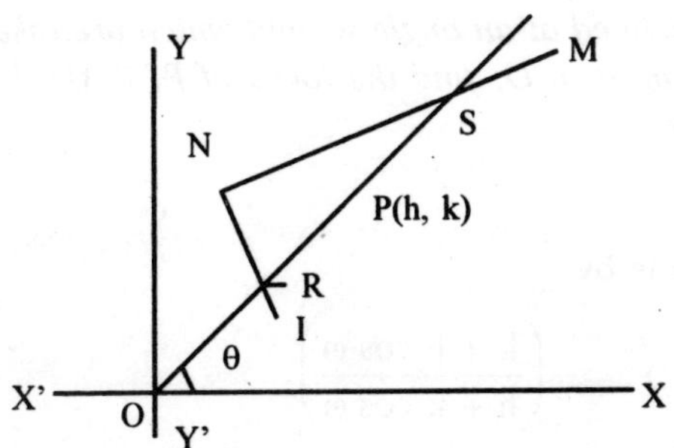

Changing again to cartesian co-ordinates and simplifying, we get

$$c_1(a_2x + b_2y) + c_2(a_1x + b_1y) + 2(a_1x + b_1y) + (a_2x + b_2y) = 0. \quad \textbf{Ans.}$$

(2) By hypothesis, $OP^2 = OR.OS$ or $r^2 = r_1 r_2$.

Hence $r^2 = \left[\dfrac{c_1}{a_1 \cos\theta + b_1 \sin\theta}\right] + \left[\dfrac{c_2}{a_2 \cos\theta + b_2 \sin\theta}\right].$

Changing into cartesian co-ordinates and simplifying, we get

$$(a_1x + b_1y)(a_2x + b_2y) + c_1c_2 = 0. \qquad \textbf{Ans.}$$

Example 100:

Find the locus of a point which moves so that the difference of its distances from two fixed straight lines at right angles is equal to its distance from a fixed straight line.

Solution:

Let the two fixed lines at right angles be the axes of x and y. Fixed given line be $x \cos\alpha + y \sin\alpha = p$ and co-ordinates of the moving point at any position be (h, k).

Its distance from x-axis is k, from y-axis is h, and from $x \cos\alpha + y \sin\alpha - p = 0$ is $(h \cos\alpha + k \sin\alpha - p)$.

By hypothesis, $k - h = h \cos\alpha + k \sin\alpha - p$

$$h(1 + \cos\alpha) - k(1 - \sin\alpha) - p = 0.$$

Generalising, we get $x(1 + \cos\alpha) - y(1 - \sin\alpha) = p$.

This represents a straight line. Hence the required locus a straight line.

Example 101(a):

From a point P perpendiculars PM and PN are drawn upon two fixed lines which are inclined at an angle w, and which are taken as the axes of co-ordinates and meet in O; find the locus of P, if MN be parallel to the given line y = mx.

Solution:

Slope of MN is by

$$-\left(\frac{k + h\cos\omega}{h + k\cos\omega}\right)$$

By hypothesis, MN is parallel to y = mx. Hence

$$m = -\left[\frac{k + h\cos\omega}{h + k\cos\omega}\right]$$

Simplifying and generalising, we get

$$x\,(m + \cos\omega) + y\,(1 + m\cos\omega) = 0. \qquad \textbf{Ans.}$$

Example 101(b):

Two fixed points A and B are taken on the axes such that OA = a and OB = b; two variable points A' and B' are taken on the same axes; find the locus of the intersection of AB' and A'B(1) when OA' + OB' = OA + OB, and (2) when $\frac{1}{OA'} - \frac{1}{OB'} = \frac{1}{OA} - \frac{1}{OB}$.

Solution:

Let the points A' and B' be (a', 0) and (0, b').

(1) By hypotheses is, OA' + OB' = OA + OB

or $\quad a' + b' = a + b$

or $\quad a' - a = b - b'. \qquad$ ...(1)

Equation of the lines AB' and A'B respectively are

$$\frac{x}{a} + \frac{y}{b'} = 1 \qquad ...(2)$$

and $\quad \frac{x}{a'} + \frac{y}{b} = 1. \qquad$ ...(3)

Subtracting (3) from (2), we get $x\left(\frac{1}{a} - \frac{1}{a'}\right) + y\left(\frac{1}{b'} - \frac{1}{b}\right) = 1$

or $\quad \frac{x}{aa'}(a' - a) + \frac{y}{bb'}(b - b') = 0$ or $\frac{x}{aa'} + \frac{y}{bb'} = 0 \qquad$ ...(4)

[∵ a' – a = b – b' (1)].

Again dividing (2) by b, (3) by a and adding, we get

$$\frac{x}{ab} + \frac{y}{bb'} + \frac{x}{a'a} + \frac{y}{ab} = \frac{1}{a} + \frac{1}{b}$$

or $\quad \frac{x}{ab} + \frac{y}{ab} = \frac{b+a}{ab} \qquad \left[\because \frac{x}{aa'} + \frac{y}{bb'} = 0 \text{ by } (4)\right]$

or $\quad x + y = a + b. \qquad$ **Ans.**

(2) By hypothesis, $\frac{1}{OA'} - \frac{1}{OB'} = \frac{1}{OA} - \frac{1}{OB}$.

or $\quad \frac{1}{a'} - \frac{1}{b'} = \frac{1}{a} - \frac{1}{b}$

or $$\left(\frac{1}{a'}-\frac{1}{a}\right)=\left(\frac{1}{b'}-\frac{1}{b}\right). \qquad ...(5)$$

Again subtracting (3) from (2), we get $x\left(\frac{1}{a}-\frac{1}{a'}\right)+y\left(\frac{1}{b'}-\frac{1}{b}\right)=0$

or $$y = x\left[\text{as by (5), }\left(\frac{1}{a'}-\frac{1}{a}\right)=\left(\frac{1}{b'}-\frac{1}{b}\right)\right].$$ **Ans.**

Example 101(c):

If a triangle ABC remain always similar to a given triangle, and if the point A be fixed and the point B always move along a given straight line, find the locus of the point C.

Solution:

Let us suppose the fixed point A as origin. The line perpendicular to the fixed line be the initial line.

Δ ABC is always similar to the given triangle. Hence

$$\frac{AB}{AC} = \text{constant} = \lambda \text{ (say)}.$$

Therefore c = AB cos (θ – A) = λr cos (θ – A)

This is the required equation which is straight line.

Example 102:

A right-angled triangle ABC, having C a right angle, is of given magnitude, and the angular points A and B slide along two given perpendicular axes; show that the locus of C is the pair of straight line whose equations are y = ± b/ax.

Solution:

Let the points A, B and P be respectively (h , 0), (0, k) and (x, y). Then

$$BP^2 = x^2 + (y-k)^2 \qquad ...(1)$$

and $$AP^2 = (x-h)^2 + y^2 \qquad ...(2)$$

Again slopes of BP and AP are respectively $\frac{y-k}{x-0}$ and $\frac{y-0}{x-h}$

As they are perpendicular to each other, $\frac{y-k}{x}\times\frac{y}{x-h}=-1$

Squaring, we get $$\frac{(y-k)^2\, y^2}{x^2\,(x-h)^2}=1 \qquad ...(3)$$

$\because$ BP = a, AB = b, by (1) and (2), we get

$$(y - k)^2 = (a^2 - x^2) \text{ and } (x - h)^2 = (b^2 - y^2).$$

Putting the values, (3) becomes $\dfrac{(a^2 - x^2)\, y^2}{x^2\,(b^2 - y^2)} = 1$

or $\quad a^2y^2 - x^2y^2 = b^2x^2 - x^2y^2$

or $\quad \dfrac{y^2}{x^2} = \dfrac{a^2}{b^2}$ or $\dfrac{y}{x} = \pm\dfrac{a}{b}$. **Proved.**

Example 103(a):

From a point P perpendiculars PM and PN are drawn upon two fixed lines which are inclined at an angle ω, and which are taken as the axes of co-ordinates and meet in O; find the locus of P, if OM – ON be equal to 2d.

Solution:

By hypothesis; OM – ON = 2d

or $\quad x + y\cos\omega - y\, x\cos\omega = 2d$

or $\quad x\,(1 - \cos\omega) - y\,(1 - \cos\omega) = 2d$

or $\quad x - y = d.\operatorname{cosec}^2\left(\dfrac{\omega}{2}\right)$ **Ans.**

Example 103(b):

From a point P perpendicular PM and PN are drawn upon two fixed lines which are inclined at an angle ω, and which are taken as the axes of co-ordinates and meet in O; find the locus of P, if PM + PN be equal to 2c.

Solution:

By hypothesis, PM + PN = 2c

or $\quad y\sin\omega + x\sin\omega = 2c$ or $x + y = 2c\operatorname{cosec}\omega$. **Ans.**

Example 103(c):

From a point P perpendicular PM and PN are drawn upon two fixed lines which are inclined at an angle ω, and which are taken as the axes of co-ordinates and meet in O; find the locus of P, if PM – PN be equal to 2c.

Solution:

By hypothesis, PM – PN = 2c

or $\quad y \sin\omega - x \sin\omega = 2c$

or $\quad x - y = 2c \operatorname{cosec}\omega.$ **Ans.**

Example 104(a):

From a point P perpendicular PM and PN are drawn upon two fixed lines which are inclined at an angle ω, and which are taken as the axes of co-ordinates and meet in O; find the locus of P, if MN be equal to 2c.

Solution:

By hypothesis, $MN = 2c$ or $MN^2 = 4c^2$

or $\quad PM^2 + PN^2 - 2.PM.PN \cos NPM = 4c^2$ (from ΔPMN)

or $\quad y^2 \sin^2\omega + x^2 \sin^2\omega = 2.y \sin\omega.x \sin\omega. \cos(180° - w) = 4c^2.$

[∴ in quadratic OMPN, ∠ OMP = ∠ ONP = 90° and ∠ MON = ω]

or $\quad y^2 + x^2 + 2xy \cos\omega = 4c^2 \operatorname{cosec}^2\omega.$ **Ans.**

Example 105(b):

From a point P perpendicular PM and PN are drawn upon two fixed lines which are inclined at an angle ω, and which are taken as the axes of co-ordinates and meet in O; find the locus of P, if MN pass through the fixed point (a, b).

Solution:

Let the co-ordinates of P be (h, k)

Then the co-ordinates of M and respectively will be $(h + k \cos\omega, 0)$ and $(0, k + h \cos\omega)$.

Then the equation of MN will be

$$\frac{x}{h + k\cos\omega} + \frac{y}{k + h\cos\omega} = 1 \qquad ...(1)$$

If the line passes through (a, b), point will satisfy it; hence

$$\frac{a}{h + k\cos\omega} + \frac{b}{k + h\cos\omega} = 1$$

Generalising, we get $\dfrac{a}{x + y\cos\omega} + \dfrac{b}{y + x\cos\omega} = 1$

On simplifying, we get

$$x(a\cos\omega + b) + y(a + b\cos\omega)$$

$$= (x^2 + y^2)\cos\omega + xy(1 - \cos^2\omega)$$ **Ans.**

Example 106:

Two given straight lines meet in O, and through a given point P is drawn a straight line to meet them in Q and R; if the parallelogram OQSR be completed find the equation to the locus of S.

Solution:

Let the axes be along OQ and OR and the co-ordinates of Q and R be respectively (a, 0) and (0, b).

Then the equation of the line RQ is $\frac{x}{a} + \frac{y}{b} = 1$

As the line passes through a fixed point P, whose co-ordinates are (x_1, y_2) say, then it will satisfy the equation (1); hence

$$\frac{x_1}{a} + \frac{y_1}{b} = 1 \qquad ...(2)$$

OQSR is parallelogram. Let the co-ordinates of S be (h, k). Then it is clear that $h = a$ ...(3)

and $k = b$ ...(4)

Substituting the values of a and b in (2), we get $\frac{x_1}{h} + \frac{y_1}{k} = 1$

Generalising, we get the locus of S as $\frac{x_1}{x} + \frac{y_1}{y} = 1$ **Ans.**

Example 107:

The base BC (=2a) of a triangle ABC is fixed; the axes being BC and a perpendicular to it through its middle point, find the locus of the vertex A, when the product of the tangents of the base angles is given (=λ).

Solution:

By hypothesis, $\tan\theta_1 \times \tan\theta_2 = \lambda$

Hence $$\left(\frac{y}{x+a}\right) \times \left(-\frac{y}{x-a}\right) = \lambda$$

or $$\lambda x^2 + y^2 = \lambda a^2.$$ **Ans.**

Example 108:

The base BC (=2a) of a triangle ABC is fixed; the axes being BC and a perpendicular to it through its middle point, find the locus of the vertex A, when the product of the tangents of the base angles is in times the tangent of the other.

Solution:

By hypothesis, $\tan\theta_2 = m\tan\theta_1$

or $$-\frac{y}{x-a} = m\frac{y}{x+a}$$

or $(m+1)x = (m-1)a$. **Ans.**

Example 109:

The base BC (=2a) of a triangle ABC is fixed; the axes being BC and a perpendicular to it through its middle point, find the locus of the vertec A, when the m times the square of one side added to n times the square of the other side is equal to a constant quantity c^2.

Solution:

By hypothesis, $nAB^2 + m.AC^2 = c^2$.

or $n[(x+a)^2 + (y-0)^2] + m[(x-a)^2 + (y-0)^2] = c^2$

or $(m+n)(x^2+y^2+a^2) - 2ax(m-n) = c^2$. **Ans.**

Example 110(a):

The base BC (=2a) of a triangle ABC is fixed; the axes being BC and a perpendicular to it through its middle point, find the locus of the vertec A, when the difference of the base angles is given (=α).

Solution:

Let the co-ordinates of A be (x, y) and $\angle ABC = \theta_1$ and $\angle ACB = \theta_2$. Then

$$\tan\theta_1 = \frac{y-0}{x+a} \text{ and } \tan(180-\theta_2) = \frac{y-0}{x-a}$$

[∴ slope of AC will be $(180-\theta_2)$]

$$\therefore \quad \tan\theta_1 = \frac{y}{x+a} \text{ and } \tan\theta_2 = -\frac{y}{x-a}$$

By hypothesis, $\theta_2 - \theta_1 = \alpha$ or $\tan(\theta_2-\theta_1) = \tan\alpha$

or $$\frac{\tan\theta_2 - \tan\theta_1}{1+\tan\theta_2\tan\theta_1} = \tan\alpha$$

or $$\frac{-\frac{y}{x-a} - \frac{y}{x+a}}{1 - \frac{y}{x-a}\frac{y}{x+a}} = \tan\alpha$$

or $$\frac{-2xy}{x^2 - a^2 - y^2} = \frac{1}{\cot \alpha}$$

or $$x^2 + 2xy \cot \alpha - y^2 = a^2.$$

Example 110(b):

Through a given point O is drawn a straight line to meet two given parallel straight lines in P and Q; through P and Q are drawn straight lines in given directions to meet in R; prove that the locus of R is a straight line.

Solution:

Let O be the origin, a line parallel to the given line be y-axis, and let the distances of the parallel lines from origin be a and b. Let the moving point R be (x, y). If $\angle$ POX = θ, the point P will be (a, a tan θ) and the co-ordinates of Q will be (b, b tan θ).

The equation of PR will be, $y - a \tan \theta = (x - a) \tan \alpha$. ...(1)

The equation of RQ will be, $y - b \tan \theta = (x - b) \tan \beta$, ...(2)

where α and β are the given angles for direction.

Multiplying (1) by b and (2) by a, and subtracting, we get

$$by - ay = bx \tan \alpha - ax \tan \beta + ab \tan \beta$$

or $$y (b - a) = x (b \tan \alpha - a \tan \beta) + ab (\tan \beta - \tan \alpha)$$

which is the required locus, and it represents a straight line. **Proved.**

Example 111:

Given n straight lines and a fixed point O; through O is drawn a straight line meeting these lines in the points R_1, R_2, R_3, ...R_n, and on it is taken a point R such that $\frac{n}{OR} = \frac{1}{OR_1} + \frac{1}{OR_2} + \frac{1}{OR_3} + \dots + \frac{1}{OR_n}$; *show that the locus of R is a straight line.*

Solution:

Let the equations of the straight lines be $a_1x + b_1y + c_1 = 0$, $a_2x + b_2y + c_2 = 0$, and so on, their polar equations will be

$$\frac{1}{r_1} = \frac{a_1 \cos \theta + b_1 \sin \theta}{c_1}$$

and so on. If the line makes an angle θ with initial line (x-axis); then by hypothesis, we have

$$\frac{n}{r} = \frac{a_1 \cos\theta + b_1 \sin\theta}{c_1} + \frac{a_2 \cos\theta + b_2 \sin\theta}{c_1} + \ldots.$$

or $$n = r\cos\theta\left[\frac{a_1}{c_1} + \frac{a_2}{c_2} + \ldots\right] + r\sin\theta\left[\frac{b_1}{c_1} + \frac{b_2}{c_2} + \ldots\right].$$

or $$n = x\sum\frac{a_1}{c_1} + x\sum\frac{b_1}{c_1},$$ which is a straight line. **Ans.**

Example 112:

From a point P perpendiculars PM and PN are drawn upon two fixed lines which are inclined at an angle ω, and which are taken as the axes of co-ordinates and meet in O; find the locus of P if OM + ON be equal to 2C.

Solution:

Let the co-ordinates of P be (x, y). Then by hypothesis

$$OM + ON = 2c.$$

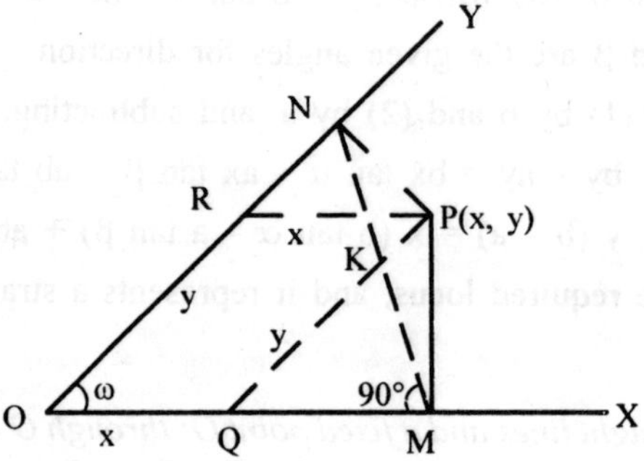

or $$x + y\cos\omega + y + x\cos\omega = 2c.$$

or $$x(1 + \cos\omega) + y(1 + \cos\omega) = 2c$$

or $$x.2\cos^2\frac{\omega}{2} + y.2\cos^2\frac{\omega}{2} = 2c$$

or $$x + y = c.\sec^2\left(\frac{\omega}{2}\right).$$ **Ans.**

Example 113:

Prove that the area of the triangle formed by the three straight lines

$$a_1x + b_1y + c_1 = 0,$$
$$a_2x + b_2y + c_2 = 0$$
and $$a_3x + b_3y + c_3 = 0 \text{ is}$$

$$\frac{1}{2}\begin{vmatrix} a_1 & b_1 & c_1 \\ a_2 & b_2 & c_2 \\ a_3 & b_3 & c_3 \end{vmatrix}^2 + (a_1b_2 - a_2b_1)(a_2b_3 - a_3b_2)(a_3b_1 - a_1b_3).$$

Solution:

The equations of the straight lines are given as

$$a_1x + b_1y + c_1 = 0 \qquad ...(1)$$

$$a_2x + b_2y + c_2 = 0 \qquad ...(2)$$

and $$a_3x + b_3y + c_3 = 0 \qquad ...(3)$$

Applying cross-multiplication to (1) and (2), we get

$$\frac{x}{b_1c_1 - b_2c_1} = \frac{y}{c_1a_2 - a_1c_2} = \frac{1}{a_1b_2 - b_1a_2}$$

or $$\frac{x}{\begin{vmatrix} b_1 & b_2 \\ c_1 & c_2 \end{vmatrix}} = \frac{y}{\begin{vmatrix} c_1 & c_2 \\ a_1 & a_2 \end{vmatrix}} = \frac{1}{\begin{vmatrix} a_1 & a_2 \\ b_1 & b_2 \end{vmatrix}}.$$

or $$x = \begin{vmatrix} b_1 & b_2 \\ c_1 & c_2 \end{vmatrix} + \begin{vmatrix} a_1 & a_2 \\ b_1 & b_2 \end{vmatrix}$$

and $$y = \begin{vmatrix} c_1 & c_2 \\ a_1 & a_2 \end{vmatrix} + \begin{vmatrix} a_1 & a_2 \\ b_1 & b_2 \end{vmatrix}$$

Let the point be (α_1, β_1).

Similarly the other two vertices will be (α_2, β_2) and (α_3, β_3).

Therefore the area of the Δ having its vertices (α_1, β_1), (α_2, β_2) and (α_3, β_3) will be (by Article 20).

$$= \frac{1}{2}\begin{vmatrix} \alpha_1 & \beta_1 & 1 \\ \alpha_2 & \beta_2 & 1 \\ \alpha_3 & \beta_3 & 1 \end{vmatrix}$$

$$= \frac{1}{2}\begin{vmatrix} a_1 & b_1 & c_1 \\ a_2 & b_2 & c_2 \\ a_3 & b_3 & c_3 \end{vmatrix}^2 + (a_1b_2 - a_2b_1)(a_2b_3 - b_2a_3) \times (a_3b_1 - b_3a_1)$$

(on substituting the values of α_1, β_1 etc. and simplifying). **Hence proved.**

Example 114:

Show that the orthocentre of the triangle formed by the three straight lines

$$y = m_1x + \frac{a}{m_1},\ y = m_2x + \frac{a}{m_2},\ \text{and } y = m_3x + \frac{a}{m_3},$$

is the point $\left\{-a,\ a\left(\frac{1}{m_1} + \frac{1}{m_2} + \frac{1}{m_3} + \frac{1}{m_1m_2m_3}\right)\right\}$

Solution:

The equations are given as

$$y = m_1x + \frac{a}{m_1}, \qquad \text{...(1)}$$

$$y = m_2x + \frac{a}{m_2}, \qquad \text{...(2)}$$

and $$y = m_3x + \frac{a}{m_3}, \qquad \text{...(3)}$$

Let these equations represent AB, BC and CA respectively. Solving them in pairs, the co-ordinates of A, B and C are respectively.

$$\left(\frac{a}{m_3m_1}, \frac{a}{m_3} + \frac{a}{m_1}\right), \left(\frac{a}{m_1m_2}, \frac{a}{m_1} + \frac{a}{m_2}\right), \text{and} \left(\frac{a}{m_2m_3}, \frac{a}{m_2} + \frac{a}{m_3}\right)$$

The equation of a line AL which is perpendicular to BC will be

$$y - \left(\frac{a}{m_3} + \frac{a}{m_1}\right) = -\frac{1}{m_2}\left(x - \frac{a}{m_3m_1}\right)$$

or $$y = -\frac{x}{m_2} + \frac{a}{m_1m_2m_3} + \frac{a}{m_1} + \frac{a}{m_3}. \qquad \text{...(4)}$$

Similarly the equation of a line BM, which is perpendicular to AC, will be

$$y = -\frac{x}{m_3} + \frac{a}{m_1m_2m_3} + \frac{a}{m_1} + \frac{a}{m_2}.$$

Solving (4) and (5), we get

$$x = -a \text{ and } y = \frac{a}{m_1} + \frac{a}{m_2} + \frac{a}{m_3} + \frac{a}{m_1m_2m_3}.$$

Hence the co-ordinates of the orthocentre will be

$$\left[-a, a\left(\frac{a}{m_1}+\frac{a}{m_2}+\frac{a}{m_3}+\frac{a}{m_1m_2m_3}\right)\right]$$ **Proved.**

Example 115:

The lines x + y + 1 = 0, x − y + 2 = 0, 4x + 2y + 3 = 0, and x + 2y − 4 = 0, are the equations to the sides of a quadrilateral taken in order; find the equations to its three diagonals and the equation to the line and which their middle points lie.

Solution:

Let the lines

$$x + y + 1 = 0, \qquad ...(1)$$

$$x - y + 2 = 0, \qquad ...(2)$$

$$4x + 2y + 3 = 0, \qquad ...(3)$$

and $$x + 2y - 4 = 0, \qquad ...(4)$$

represent AB, BC, CD and DA respectively.

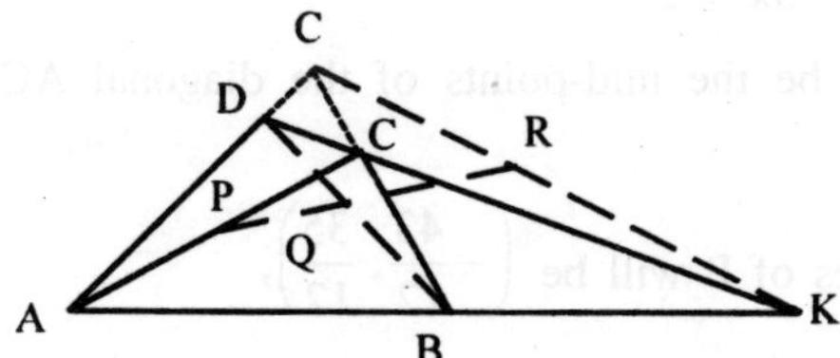

Solving (1) and (2), co-ordinates of B are $\left(-\frac{3}{2}, \frac{1}{2}\right)$.

Solving (1) and (3), co-ordinates of K are $\left(-\frac{1}{2}, -\frac{1}{2}\right)$.

Solving (1) and (4), co-ordinates of A are (–6, 5).

Solving (2) and (3), co-ordinates of C are $\left(-\frac{7}{6}, \frac{5}{6}\right)$.

Solving (2) and (4), co-ordinates of L are (0, 2).

Solving (3) and (4), co-ordinates of D are $\left(-\frac{7}{3}, \frac{19}{6}\right)$.

Equation of the diagonal AC will be

$$y - 5 = \frac{\frac{5}{6} - 5}{-\frac{7}{6} + 6}(x + 6)$$

or $\quad 25x + 29y + 5 = 0.$ ...(5)

Equation of the diagonal BD will be

$$y - \frac{1}{2} = \frac{\frac{19}{6} - \frac{1}{2}}{-\frac{7}{3} + \frac{3}{2}}\left(x + \frac{3}{2}\right)$$

or $\quad 32x + 10y + 43 = 0.$ ...(6)

Equation of the diagonal KL will be

$$y + \frac{1}{2} = \frac{2 + \frac{1}{2}}{0 + \frac{1}{2}}\left(x + \frac{1}{2}\right)$$

or $\quad y = 5x + 2.$ ...(7)

If P, Q and R be the mid-points of the diagonal AC, BD and KL respectively, then

The co-ordinates of P will be $\left(-\frac{43}{12}, \frac{35}{12}\right)$,

The co-ordinates of Q will be $\left(-\frac{23}{12}, \frac{11}{6}\right)$,

The co-ordinates of R will be $\left(-\frac{1}{4}, \frac{3}{4}\right)$.

The equation of the line joining P and Q is

$$y - \frac{11}{6} = \frac{\frac{35}{12} - \frac{11}{6}}{-\frac{43}{12} + \frac{23}{12}}\left(x + \frac{23}{12}\right)$$

or $\quad 52x + 80y = 47.$ ...(8)

Substituting the co-ordinates of R in equation (8), we find that the equation is satisfied. Hence P, Q, R are collinear.

Equations 5, 6 and 7 represent the 3 diagonals an 8 represents the line joining mid points of the diagonals.

Example 116:

A and B are two fixed points whose co-ordinates are (3, 2) and (5, 1) respectively; ABP is an equilateral triangle on the side of AB remote from the origin. Find the co-ordinates of P and the orthocentre of the triangle ABP.

Solution:

Let the co-ordinates of P be (x, y). A and B are given as (3, 2) and (5, 1) respectively. As ABP is an equilateral triangle,

$$PA = PB = AB$$

or $$PA^2 = PB^2 = AB^2.$$

Hence

$$(x-3)^2 + (y-2)^2 = (x-5)^2 + (y-1)^2 = (5-3)^2 + (1-2)^2.$$

Therefore, $x^2 + y^2 - 6x - 4y + 13 = x^2 + y^2 - 10x - 2y + 26 = 5.$

Taking in pairs,

$$x^2 + y^2 - 6x - 4y + 13 = x^2 + y^2 - 10x - 2y + 26$$

or $4x - 2y - 13 = 0$ ...(1)

and $x^2 + y^2 - 6x - 4y + 13 = 5.$ ...(2)

Substituting the value of y from (1) in (2), we get

$$x^2 \left(2x - \frac{13}{2}\right)^2 - 6x - 4\left(2x - \frac{13}{2}\right) + 13 = 5$$

or $$4x^2 - 32x - 61 = 0.$$

$$\therefore \quad x = \frac{32 \pm \sqrt{(1024 - 976)}}{2 \times 4} = 4 \pm \frac{1}{2}\sqrt{3}.$$

If point P is such as to be remote from the origin, we have to make +ve sign only. Hence $x = 4 + 1/2\sqrt{3}$.

Putting in (1), $y = \frac{3}{2} + \sqrt{3}$.

Hence the required point is $\left(4 + \frac{1}{2}\sqrt{3}, \frac{3}{2} + \sqrt{3}\right)$. **Ans.**

Equation of the line PL which is perpendicular to AB will be

$$y - \left(\frac{3}{2} + \sqrt{3}\right) = \left[x - \left(4 + \frac{1}{2}\sqrt{3}\right)\right] \quad \text{... 'm' of AB is } \frac{2-1}{3-5} = -\frac{1}{2}$$

or $$y = 2x - \frac{13}{2}. \qquad ...(3)$$

Similarly equation of AM, the perpendicular on PB from A, is

$$y - 2 = \frac{5\sqrt{3} - 8}{11}(x - 3).$$

or $$11y - (5\sqrt{3} - 8)x = 46 - 15\sqrt{3}. \qquad ...(4)$$

Solving (3) and (4), we get

$$x = 4 + \frac{1}{6}\sqrt{3}$$

and $$y = \frac{3}{2} + \frac{1}{3}\sqrt{3}.$$

Hence the co-ordinates of the orthocentre are

$$\left(4 + \frac{1}{6}\sqrt{3}, \frac{3}{2} + \frac{1}{3}\sqrt{3}\right).$$

Example 117:

The vertices of a quadrilateral, taken in order, are the points (0, 0), (4, 0), (6, 7) and (0, 3); find the co-ordinates of the points of intersection of the two lines joining the middle points of opposite sides.

Solution:

Let the given co-ordinates (0, 0), (4, 0), (6, 7) and (0, 3) represent A, B, C and D respectively. If P, Q, R and S be the mid-point of AB, BC, CD and DA respectively, then their co-ordinates are respectively (2, 0), (5, 3/2) (3, 5) and (0, 3/2).

As by plane geometry, the lines joining the mid-points of the opposite sides of any quadrilateral bisect each other, if M is the point of intersection, it is the mid-point or PR.

The co-ordinates of the mid-point of P and R will be $\left(\frac{5}{2}, \frac{5}{2}\right)$.

Hence the required point is $\left(\frac{5}{2}, \frac{5}{2}\right)$. **Ans.**

Example 118(a):

Find the area of the triangle formed by the straight line whose equations are $3x - 4y + 4a = 0$, $2x - 3y + 4a = 0$, and $5x - y + a = 0$, proving also that the feet of the perpendiculars from the origin upon them are collinear.

Solution:

The equation of the lines are given as

$$3x - 4y + 4a = 0 \quad \ldots(1)$$

$$2x - 3y + 4a = 0 \quad \ldots(2)$$

and $$5x - y + a = 0 \quad \ldots(3)$$

Salving them by two, the co-ordinates of the vertices are (4a, 4a), (0, a) and (a/13, 18a/13).

$\therefore$ Area of the triangle

$$= \frac{1}{2}\left[4a\left(a - \frac{18a}{13}\right) + 0\left(\frac{18a}{13} - 4a\right) + \frac{a}{13}(4a - a)\right] = \frac{17a^2}{26} \text{ units. } \textbf{Ans.}$$

Again, the equation of any line perpendicular to (1) is

$$4x + 3y = \lambda, \quad \ldots(4)$$

and as it passes through origin, (0, 0) will satisfy (4); hence l = 0.

$\therefore$ Equation of the perpendicular from origin on (1) is

$$4x + 3y = 0. \quad \ldots(5)$$

Similarly the equation of the perpendicular from origin on (2) will be

$$3x + 2y = 0 \quad \ldots(6)$$

and the perpendicular from origin on (3) will be

$$x + 5y = 0. \quad \ldots(7)$$

Solving (1) and (5), co-ordinates of the foot of the perpendicular

$$= \left(-\frac{12a}{25}, \frac{16a}{25}\right).$$

Solving (2) and (6), the co-ordinates of the foot of the perpendicular

$$= \left(-\frac{8a}{13}, \frac{12a}{13}\right).$$

Solving (3) and (7), the co-ordinates of the foot of the perpendicular

$$= \left(-\frac{5a}{26}, \frac{a}{26}\right).$$

Hence the area of thc triangle foimed by joining the feet of perpendiculars.

$$= \frac{1}{2}\left[-\frac{12a}{25}\left(\frac{12a}{13} - \frac{a}{26}\right) - \frac{8a}{13}\left(\frac{a}{26} - \frac{16a}{25}\right) - \frac{5a}{26}\left(\frac{16a}{25} - \frac{12a}{13}\right)\right]$$

= 0 (on simplification).

Hence the three points are collinear.

Example 118(b):

Prove that the area of the triangle formed by the three straight lines $x \cos\alpha + y \sin\alpha - p_1 = 0$, $x \cos\beta + y \sin\beta - p_2 = 0$ *and* $x \cos\gamma + y \sin\gamma - p_3 = 0$ *is*

$$\frac{1}{2}\frac{\{p_1 \sin(\gamma-\beta) + p_2 \sin(\alpha-\gamma) + p_3 \sin(\beta-\alpha)\}^2}{\sin(\gamma-\beta)\sin(\alpha-\gamma)\sin(\beta-\alpha)}.$$

Solution:

The lines are given as

$$x \cos\alpha + y \sin\alpha - p_1 = 0 \quad \text{...(1)}$$

$$x \cos\beta + y \sin\beta - p_2 = 0 \quad \text{...(2)}$$

and $$x \cos\gamma + y \sin\gamma - p_3 = 0 \quad \text{...(3)}$$

Solving the given equations in pairs, we get the co-ordinates of the vertices as

$$\left[\frac{p_1 \sin\beta - p_2 \sin\alpha}{\sin(\beta-\alpha)}, \frac{p_2 \cos\alpha - p_1 \cos\beta}{\sin(\beta-\alpha)}\right]$$

$$\left[\frac{p_2 \sin\gamma - p_3 \sin\beta}{\sin(\gamma-\beta)}, \frac{p_3 \cos\beta - p_2 \cos\gamma}{\sin(\gamma-\beta)}\right]$$

and $$\left[\frac{p_3 \sin\alpha - p_1 \sin\gamma}{\sin(\alpha-\gamma)}, \frac{p_1 \cos\gamma - p_3 \cos\alpha}{\sin(\alpha-\gamma)}\right]$$

Therefore the area of the triangle

$$= \frac{1}{2}\frac{\{p_1 \sin(\gamma-\beta) + p_2 \sin(\alpha-\gamma) + p_3 \sin(\beta-\alpha)\}^2}{\sin(\gamma-\beta)\sin(\alpha-\gamma)\sin(\beta-\alpha)}$$

[on putting the value and simplifying]. **Proved.**

Example 119:

Find the area of the triangle formed by the straight lines whose equations are $y = m_1x + c_1$, $y = m_2x + c_2$, *and the axis of y.*

Solution:

The equations of the straight lines are given as

$$y = m_1x + c_1, \quad \ldots(1)$$

$$y = m_2x + c_2, \quad \ldots(2)$$

and $$x = 0 \quad \ldots(3)$$

Solving in pairs the co-ordinates of the vertices are

$$\left(\frac{c_2 - c_1}{m_1 - m_2}, \frac{m_1c_1 - m_2c_2}{m_1 - m_2}\right), (0, c_1) \text{ and } (0, c_1).$$

$$\therefore \quad \text{Area of } \Delta = \frac{1}{2}\left[\frac{c_2 - c_1}{m_1 - m_2}(c_2 - c_1)\right] = \frac{1}{2}\frac{(c_2 - c_1)^2}{(m_1 - m_2)}. \qquad \textbf{Ans.}$$

Example 120:

Find the area of the triangle formed by the straight lines whose equations are $y = m_1x + c_1$, $y = m_2x + c_2$, *and* $y = m_3x + c_3$.

Solution:

The equation of th straight lines are given as

$$y = m_1x + c_1, \quad \ldots(1)$$

$$y = m_2x + c_2, \quad \ldots(2)$$

and $$y = m_3x + c_3, \quad \ldots(3)$$

Taking any two straight lines and y-axis, the area of the triangle may be found as in Q. No. 207. Taking the points in order, we find that the area of the triangle formed by (1), (2) and (3) is the sum of the three triangles formed by any two straight lines and y-axis. Hence total area will be

$$\frac{1}{2}\left[\frac{(c_2 - c_1)^2}{m_1 - m_2} + \frac{(c_2 - c_3)^2}{m_2 - m_3} + \frac{(c_3 - c_1)^2}{m_3 - m_1}\right]. \qquad \textbf{Ans.}$$

Example 121(a):

Find the area of the triangle formed by the straight lines whose equations are $y = ax - bc$, $y = bx - ca$, *and* $y = cx - ab$.

Solution:

The equations of the straight lines are

$$y = ax - bc \quad ...(1)$$

$$y = bx - ca \quad ...(2)$$

and $$y = cx - ab \quad ...(3)$$

Solving them in pairs, the co-ordinates of the vertices are

$$\{-c, -c(a + b)\},$$

$$\{-a, -a(b + c)\}$$

and $$\{-b, -b(a + c)\}.$$

Area of the triangle

$$= \frac{1}{2}[-c\{-a(b + c) + b(a + c)\} - a\{-b(a + c) + c(a + b)\} - b\{-c(a + b) + a(b + c)\}]$$

$$= \frac{1}{2}[-c\{-ac + cb\} - a\{-ba + ca\} - b\{-cb + ab\}]$$

$$= \frac{1}{2}[ac^2 - bc^2 + ba^2 - ca^2 + cb^2 - ab^2]$$

Factorizing and omitting the –ve sign, we get

$$= \frac{1}{2}(a - b)(b - c)(c - a).$$ **Ans.**

Example 122(b):

Find the area of the triangle formed by the straight lines whose equations are

$$2y + x - 5 = 0,$$

$$y + 2x - 7 = 0$$

and $$x - y + 1 = 0.$$

Solution:

The equations of the lines are

$$2y + x - 5 = 0, \quad ...(1)$$

$$y + 2x - 7 = 0 \quad ...(2)$$

and $$x - y + 1 = 0. \quad ...(3)$$

Solving them the pairs, the co-ordinates of the angular points are (3, 1), (2, 3) and (1, 2). Therefore, required are

$$= \frac{1}{2}[3(3-2) + 2(2-1) + 1(1-3)] = \frac{3}{2} \text{ units. } \textbf{Ans.}$$

Example 123:

Prove that the area of the parallelogram whose sides are the straight lines

$$a_1x + b_1y + c_1 = 0,$$
$$a_1x + b_1y + d_1 = 0,$$
$$a_2x + b_2y + c_2 = 0,$$

and $a_2x + b_2y + d_2 = 0$ *is* $\dfrac{(d_1 - c_1)(d_2 - c_2)}{a_1b_2 - a_2b_1}$.

Solution:

Let ABCD be any parallelogram. If AL and AM be perpendiculars from A on BC and CD respectively, then

BC = AD = AM cosec θ, when θ is the angle ABC.

Area of the parallelogram = BC.AL = AM.AL cosec θ. ...(1)

Let the equations of the lines AD and BC be

$$a_1x + b_1y + c_1 = 0$$

and $$a_1x + b_1y + d_1 = 0.$$

Perpendicular distance between these two lines = AL = $\dfrac{d_1}{\sqrt{(a_1^2 + b_1^2)}}$. ...(2)

Similarly, If AB and CD be given $a_2x + b_2y + c_2 = 0$ and $a_2x + b_2y + d_2 = 0$, the distance between them,

i.e., $$AM = \frac{d_2 - c_2}{\sqrt{(a_2^2 + b_2^2)}}. \quad ...(3)$$

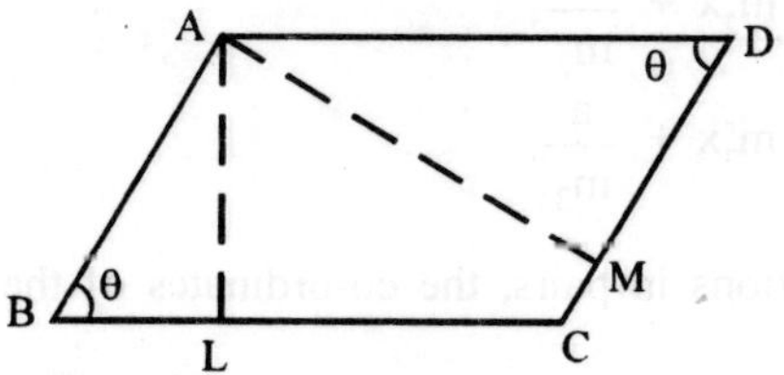

As θ is the angle between the sides AB and BC,

$$\tan\theta = \frac{b_1a_2 - a_1b_2}{a_1a_2 + b_1b_2} \text{ (by Article 66).}$$

$$\therefore \quad \operatorname{cosec}\theta = \sqrt{\left(\frac{1+\tan^2\theta}{\tan^2\theta}\right)} = \frac{\sqrt{\left(a_1^2+b_1^2\right)\left(a_2^2+b_2^2\right)}}{b_1a_2 - a_1b_2} \qquad \text{...(4)}$$

Substituting the values of (2), (3) and (4), we get

$$\text{Required area} = \frac{\left(d_1 - c_1\right)}{\sqrt{\left(a_1^2+b_1^2\right)}} \cdot \frac{\left(d_2 - c_2\right)}{\sqrt{\left(a_2^2+b_2^2\right)}} \cdot \frac{\sqrt{\left\{\left(a_1^2+b_1^2\right)\left(a_2^2+b_2^2\right)\right\}}}{b_1a_2 - a_1b_2}$$

$$= \frac{\left(d_1 - c_1\right)\left(d_2 - c_2\right)}{a_1b_2 - a_2b_1} \text{ (omitting –ve sign).} \qquad \textbf{Proved.}$$

Example 124:

Find the area of the triangle formed by the straight lines whose equations are

$$y = m_1x + \frac{a}{m_1},$$

$$y = m_2x + \frac{a}{m_2},$$

and $$y = m_3x + \frac{a}{m_3}.$$

Solution:

The equations of the given lines are

$$y = m_1x + \frac{a}{m_1}, \qquad \text{...(1)}$$

$$y = m_2x + \frac{a}{m_2}, \qquad \text{...(2)}$$

and $$y = m_3x + \frac{a}{m_3}. \qquad \text{...(3)}$$

Solving the equations in pairs, the co-ordinates of the vertices are

$$\left\{\frac{a}{m_1m_2}, \left(\frac{a}{m_1}+\frac{a}{m_2}\right)\right\}, \left\{\frac{a}{m_2m_3}, \left(\frac{a}{m_2}+\frac{a}{m_3}\right)\right\}$$

$$\text{and } \left\{\frac{a}{m_3m_1}, \left(\frac{a}{m_3}+\frac{a}{m_1}\right)\right\}$$

Hence the area of the triangle

$$= \frac{1}{2}\left[\frac{a}{m_1m_2}\left(\frac{a}{m_2}+\frac{a}{m_3}-\frac{a}{m_3}-\frac{a}{m_1}\right)+\frac{a}{m_2m_3}\left(\frac{a}{m_3}+\frac{a}{m_1}-\frac{a}{m_1}-\frac{a}{m_2}\right)\right.$$

$$\left.+\frac{a}{m_3m_1}\left(\frac{a}{m_1}+\frac{a}{m_2}-\frac{a}{m_2}-\frac{a}{m_3}\right)\right]$$

$$= \frac{1}{2}a^2\left[\frac{1}{m_1m_2^2}-\frac{1}{m_1^2m_2}+\frac{1}{m_2m_3^2}-\frac{1}{m_2^2m_3}+\frac{1}{m_3m_1^2}-\frac{1}{m_3^2m_1}\right]$$

$$= \frac{a^2}{2}\left(\frac{1}{m_1}-\frac{1}{m_2}\right)\left(\frac{1}{m_2}-\frac{1}{m_3}\right)\left(\frac{1}{m_3}-\frac{1}{m_1}\right)$$ **Ans.**

Example 125:

Prove that the area of the parallelogram contained by the lines $4y - 3x - a = 0$, $3y - 4x + a = 0$, $4y - 3x - 3a = 0$, and $3y - 4x + 2a = 0$ is $2/7\ a^2$.

Solution:

The equations of the sides of the parallelogram are

$4y - 3x - a = 0,$ ...(1)

$3y - 4x + a = 0,$ ...(2)

$4y - 3x - 3a = 0,$ ...(3)

and $3y - 4x + 2a = 0.$ ...(4)

The area of the parallelogram will be twice the area of the triangle formed by the line joining any 3 vertices of it.

Solving (1) and (2), we get the co-ordinates of one vertex as = (a, a).

Solving (1) and (4), we get the co-ordinates of other vertex $= \left(\frac{11a}{7}, \frac{10a}{7}\right)$.

Solving (2) and (3), we get the o-rdinates of third vertex $= \left(\frac{13a}{7}, \frac{15a}{7}\right)$.

$\therefore$ Area of the parallelogram

$$= 2 \times \frac{1}{2}\left[a\left(\frac{10a}{7} - \frac{15a}{7}\right) + \frac{11a}{7}\left(\frac{15a}{7} - a\right) + \frac{13a}{7}\left(a - \frac{10a}{7}\right)\right]$$

$$= a^2\left[-\frac{5}{7} + \frac{11}{7}.\frac{8}{7} - \frac{13}{7}.\frac{3}{7}\right] = \frac{2a^2}{7}. \qquad \textbf{Ans.}$$

Example 126:

Find the area of the triangle formed by the straight lines whose equations are y = x, y = 2x, and y = 3x + 4.

Solution:

The equations of the straight lines are given as

$$y = x \qquad ...(1)$$

$$y = 2x \qquad ...(2)$$

and $$y = 3x + 4. \qquad ...(3)$$

Solving (1) and (2), we get the co-ordinates of a vertex as (0, 0).

Solving (2) and (3), we get the co-ordinate of second vertex as (–2, –2).

Solving (2) and (3), we get the co-ordinates f 3rd vertex as (–4, –8).

Hence area of the triangle

$$= \frac{1}{2}\ [0\ (-2 + 4) - 2\ (-8 - 0) - 4\ (0 + 2)] = 4 \text{ units.} \qquad \textbf{Ans.}$$

Example 127:

Find the area of the triangle formed by the straight lines whose equations are

$$y + x = 0,$$

$$y = x + 6,$$

and $$y = 7x + 5.$$

Solution:

The equations of the straight lines are given as

$$y + x = 0 \qquad ...(1)$$

$$y = x + 6 \qquad ...(2)$$

$$y = 7x + 5 \qquad ...(3)$$

Co-ordinates of the point of intersection of (1) and (2) are (–3, 3).

Co-ordinates of the point of intersection of (2) and (3) are $\left(\frac{1}{6}, \frac{37}{6}\right)$.

Co-ordinates of the point of intersection of (1) and (3) are $\left(-\frac{5}{8}, \frac{5}{8}\right)$.

Area of the triangle $= \frac{1}{2}\left[-3\left(\frac{37}{6} - \frac{5}{8}\right) + \frac{1}{6}\left(\frac{5}{8} - 3\right) - \frac{5}{8}\left(3 - \frac{37}{6}\right)\right] = 7\frac{25}{48}$

(omitting sing). **Ans.**

Example 128:

Find the position of the centre of the circle circumscribing the triangle whose verities are the points

(2, 3), (3, 4) and (6, 8).

Solution:

The co-ordinates of the vertices are given as (2, 3), (3, 4) and (6, 8). Let the points be A, B and C respectively.

Equation of AB, $\quad y - 3 = \frac{4-3}{3-2}(x - 2)$

or $\quad y - 3 = x - 2. \qquad ...(1)$

The middle point of AB $= \left(\frac{2+3}{2}, \frac{3+4}{2}\right)$ or $\left(\frac{5}{2}, \frac{7}{2}\right)$.

The equation of the line perpendicular to (1) and passing through $\left(\frac{5}{2}, \frac{7}{2}\right)$ is

$$\left(y - \frac{7}{2}\right) = (-1)\left(x - \frac{5}{2}\right)$$

or $\quad x + y - 6 = 0. \qquad ...(2)$

Again, the slope of AC $= \frac{8-3}{6-2} = \frac{5}{4}$.

The co-ordinates of the mid-point of AC

$$= \left(\frac{2+6}{2}, \frac{3+8}{2}\right) \text{ or } \left(4, \frac{11}{2}\right).$$

Equation of the passing line through $\left(4, \frac{11}{2}\right)$ and perpendicular to AC will be

$$y - \frac{11}{2} = -\frac{4}{5}(x - 4).$$

or $\quad 10y - 55 = -8x + 32$

or $\quad 8x + 10y = 87$ ...(3)

Solving equations (2) and (3), we get

$$x = -\frac{27}{2},\ y = \frac{39}{2}.$$

$\therefore$ Required centre $= \left(-\frac{27}{2}, \frac{39}{2}\right)$. **Ans.**

Its radius $= \dfrac{2+12-6}{\sqrt{(1+1)}} = \dfrac{8}{\sqrt{2}} = 4\sqrt{2}$ **Ans.**

Centre of the circle opposite to vertex B is

$$= \left(\frac{6 \times 5\sqrt{2} - 0 + 7 \times 6\sqrt{2}}{5\sqrt{2} - 5\sqrt{2} + 6\sqrt{2}}, \frac{0 + 30\sqrt{2} - 42\sqrt{2}}{5\sqrt{2} + 5\sqrt{2} + 6\sqrt{2}}\right) = (12, 2).$$ **Ans.**

Its radius $= \dfrac{2+12-6}{\sqrt{(1+1)}} = \dfrac{8}{\sqrt{2}} = 4\sqrt{2}$ units. **Ans.**

Centre of the circle opposite to vertex C is

$$= \left(\frac{6 \times 5\sqrt{2} + 0 - 7 \times 6\sqrt{2}}{5\sqrt{2} + 5\sqrt{2} - 6\sqrt{2}}, \frac{0 + 30\sqrt{2} - 42\sqrt{2}}{5\sqrt{2} + 5\sqrt{2} - 6\sqrt{2}}\right) = (-3, -3).$$

Its radius $= \dfrac{-3-3-6}{\sqrt{(1+1)}} = \dfrac{12}{\sqrt{2}} = 6\sqrt{2}$ units. **Ans.**

Example 129:

Find the co-ordinates of the points of intersection of the straight lines, whose equations are

$$y = m_1 x + \frac{a}{m_1}$$

and $$y = m_2 x + \frac{a}{m_2}.$$

Solution:

$$y = m_1 x + \frac{a}{m_1}$$

or $$m^1_2 x - m_1 y + a = 0. \quad ...(1)$$

and $$y = m_2 x$$

or $$\frac{a}{m_2} \text{ or } m_2^2 x - m_2 y + a = 0. \quad ...(2)$$

By cross-multiplication,

$$\frac{x}{(-m_1)(a)-(a)(-m_2)} = \frac{y}{(a)(m_2)^2-(a)(m_1^2)} = \frac{1}{(m_1^2)(-m_2)-(-m_2)(m_2^2)}$$

or $$\frac{x}{a(m_2 - m_1)} = \frac{y}{a(m_2 - m_1)(m_2 + m_1)} = \frac{1}{m_1 m_2 (m_2 - m_1)}$$

or $$x = \frac{a}{m_1 m_2}, \; y = \frac{a(m_2 + m_1)}{m_1 m_2}$$

so point is $$\left[\frac{a}{m_1 m_2}, a\left(\frac{1}{m_1} + \frac{1}{m_2}\right)\right]$$

Example 130:

Find the co-ordinates of the points of intersection of the straight lines, whose equations are

$$2x - 3y + 5 = 0$$

and $$7x + 4y = 3.$$

Solution:

To equations are

$$2x - 3y + 5 = 0 \qquad ...(1)$$

and

$$7x + 4y - 3 = 0. \qquad ...(2)$$

By cross-multiplication,

$$\frac{x}{(-3)(-3) - (5)(4)} = \frac{y}{(5)(7) - (2)(-3)} = \frac{1}{(2)(4) - (-3)(7)}$$

or

$$\frac{x}{-11} = \frac{y}{41} = \frac{1}{29}$$

or $x = -\frac{11}{29}, y = -\frac{41}{29}$ $\therefore$ Required point $= \left(-\frac{11}{29}, \frac{41}{29}\right)$ **Ans.**

Note: Students can solve the equations by method of elimination, i.e. multiplying (1) by 7, and (2) by 2 and subtracting etc.)

Example 131:

Find the co-ordinates of the points of intersection of the straight lines, whose equations are x/a + y/b = 1 and x/b + y/a = 1.

Solution:

The equations are $\frac{x}{a} + \frac{y}{b} = 1$

or

$$bx + ay - ab = 0 \qquad ...(1)$$

and

$$\frac{x}{a} + \frac{y}{b} = 1$$

or

$$ax + by - ab = 0 \qquad ...(2)$$

By cross-multiplication

$$\frac{1}{(a)(-ab) - (-ab)b} = \frac{y}{(-ab)(a) - b(-ab)} = \frac{1}{b^2 - a^2}.$$

or

$$\frac{x}{ab(b-a)} = \frac{y}{ab(b-a)} = \frac{1}{(b-a)(b+a)}$$

or

$$x = \frac{ab}{b+a}, y = \frac{ab}{b+a}.$$

So the required point $= \left(\frac{ab}{b+a}, \frac{ab}{b+a}\right)$. **Ans.**

Example 132(a):

Find the perpendicular distance from the origin of the perpendicular from the point (1, 2) upon the straight line $x - \sqrt{3}y + 4 = 0$.

Solution:

Any line passing through the point (1, 2) may be given by

$$y - 2 = m(x - 1). \quad ...(1)$$

If (1) is perpendicular to $x - \sqrt{3}y + 4 = 0$, ...(2)

then $m \times \dfrac{1}{\sqrt{3}} = -1$ $\left[\because \text{slope of (2) is } \dfrac{1}{\sqrt{3}}\right]$

or $m = -\sqrt{3}$.

Putting this value in (1), we get the line as

$$y - 2 = \sqrt{3}(x - 1)$$

or $\sqrt{3}x + y - 2 - \sqrt{3} = 0$. ...(3)

The length of the perpendicular on (3) from (0, 0)

$$= \frac{0\sqrt{3} + 0.1 - 2 - \sqrt{3}}{\sqrt{(3+1)}} = \frac{-(2+\sqrt{3})}{2}$$

$$= \frac{1}{2}(2 + \sqrt{3}) \text{ (omitting the –ve sign)}$$ **Ans.**

Example 132(b):

Two straight lines cut the axis of x at distances a and –a and the axis of y at distances b and b' respectively; find the co-ordinates of their point of intersection.

Solution:

The first straight line cuts axis of x at a distance of a and axis of y at a distance of b; hence its equation is

$$\frac{x}{a} + \frac{y}{b} = 1$$

or $bx + ay - ab = 0$. ...(1)

The second line cuts axis of x at a distance of (–a) and axis of y at a distance of b'; so its equation is

$$\frac{x}{-a} + \frac{y}{b'} = 1$$

or $$b'x - ay + ab = 0. \qquad ...(2)$$

To solve (1) and (2) simultaneously by cross-multiplication, we have

$$\frac{x}{(a)(ab') - (-a)(-ab)} = \frac{y}{(-ab)(b') - (b)(ab')} = \frac{1}{(b)(-a) - (a)(b')}$$

or $$\frac{x}{-a^2(b-b')} = \frac{y}{-2abb'} = \frac{1}{-a(b+b')}$$

or $$x = \frac{a(b-b')}{b+b'},\ y = \frac{2bb'}{b+b'}$$

So required point is $\left(\frac{a(b-b')}{b+b'}, \frac{2bb'}{b+b'}\right)$. **Ans.**

Example 133:

Find the co-ordinates of the points of intersection of the straight lines, whose equations are x cos ϕ_1 + y sin ϕ_1 = a and x cos f_2 + y sin ϕ_2 = a.

Solution:

The equations are

$$x \cos \phi_1 + y \sin \phi_1 - a = 0 \qquad ...(1)$$

and $$x \cos \phi_2 + y \sin \phi_2 - a = 0. \qquad ...(2)$$

By cross-multiplication

$$\frac{x}{(\sin\phi_1)(-a) - (-a)(\sin\phi_2)} = \frac{y}{(-a)(\cos\phi_2) - (\cos\phi_1)(-a)}$$

$$= \frac{1}{(\cos\phi_1)(\sin\phi_2) - (\sin\phi_2)(\cos\phi_2)}$$

or $$\frac{x}{a(\sin\phi_2 - \sin\phi_1)} = \frac{y}{a(\cos\phi_1 - \cos\phi_2)} = \frac{1}{\sin(\phi_2 - \phi_1)}$$

or $$\frac{x}{a2.\cos\frac{1}{2}(\phi_2+\phi_1)\sin\frac{1}{2}(\phi_2-\phi_1)} = \frac{y}{a2.\sin\frac{1}{2}(\phi_1+\phi_2)\sin\frac{1}{2}(\phi_2-\phi_1)}$$

$$= \frac{1}{2.\sin\frac{1}{2}(\phi_2-\phi_1)\cos\frac{1}{2}(\phi_2-\phi_1)}$$

$$x = \frac{a \cos \frac{1}{2}(\phi_2 + \phi_1)}{\cos \frac{1}{2}(\phi_2 - \phi_1)}, \; y = \frac{a \sin \frac{1}{2}(\phi_1 + \phi_2)}{\cos \frac{1}{2}(\phi_1 - \phi_2)}$$

($\because \cos(-\theta) = \cos \theta$).

Hence the required point is

$$\left\{a \cos \tfrac{1}{2}(\phi_2 + \phi_1) \sec \tfrac{1}{2}(\phi_1 - \phi_2),\; a \sin \tfrac{1}{2}(\phi_1 + \phi_2) \sec \tfrac{1}{2}(\phi_1 - \phi_2)\right\}$$

Example 134(a):

Prove that the points whose co-ordinates are respectively (5, 1) (1, –1) and (11, 4) lie on a straight line, and find its intercepts on the axes.

Solution:

The points are given as (5, 1), (1, –1) and (11, 4)

The equation of the line joining (5, 1) and (1, –1) is

$$y - 1 = \frac{-1-1}{5-1}(x - 5) \text{ or } x - 2y - 3 = 0. \quad ...(1)$$

Putting the co-ordinates of the 3rd point (11, 4) in L.H.S. on (1), we get $11 - 8 - 3 = 0$. Hence it satisfies.

Therefore the points are collinear. **Proved.**

Again (1) can be written as $x - 2y = 3$, the equation

$$\frac{x}{3} - \frac{2y}{3} = 1 \text{ or } \frac{x}{3} + \frac{y}{(-3/2)} = 1.$$

Hence the intercepts on axes are 3 and –3/2. **Ans.**

Example 134(b):

Find the equations of the two straight lines drawn through the point (0, a) on which the perpendiculars let fall from the point (2a, 2a) are each of length a.

Prove also that the equation of the straight line joining the feet of these perpendiculars is y + 2x = 5a.

Solution:

Equation of any straight line passing through (0, a) may be given by

$$y - a = m(x - 0)$$

$$mx - y + a = 0 \quad ...(1)$$

Length of the perpendicular from (2a, 2a) on (1) will be

$$\frac{m.2a - 2a + a}{\sqrt{(m^2 + 1)}} = a \qquad \text{[by hypothesis]}$$

or $\quad 2m - 1 = \sqrt{(1 + m^2)}$

or $\quad 4m^2 - 4m + 1 = 1 + m^2$

or $\quad 3m^2 - 4m = 0$

or $\quad m(3m - 4) = 0.$

Hence either $m = 0$ or $m = \frac{4}{3}$.

Putting these values in (1), we get

as $-y + a = 0$ or $y = a$ (if $m = 0$)

and $\frac{4}{3}x - y + a = 0$ or $3y = 4x + 3a$ $\left(\text{if } m = \frac{4}{3}\right)$ **Ans.**

Example 135:

Find the co-ordinates of the point in which the line 2y – 3x + 7 = 0, meets the line joining the two points (6, 2) and (–8, 7). Find also the angle between them.

Solution:

The equation of the line passing through the given points (6, 2) and (–8, 7) is

$$y + 2 = \frac{7 + 2}{-8 - 6}(x - 6)$$

or $$9x + 14y - 26 = 0 \qquad ...(1)$$

The given line is $2y + 3y - 7 = 0$

$$3x - 2y - 7 = 0 \qquad ...(2)$$

Multiplying (2) by (7) and adding (1) to it,

$$30x - 75 = 0 \text{ or } x = \frac{5}{3}$$

Putting this value in (2), we get $y = \frac{1}{4}$.

Hence the point of intersection of (1) and (2) is $\left(\frac{5}{3}, \frac{1}{4}\right)$ **Ans.**

$$\text{Again, slope of (1)} = \frac{9}{14} = (m_1).$$

$$\text{slope of (2)} = \frac{3}{2} = (m_3).$$

If θ be the angle between (1) and (2),

$$\tan\theta = \frac{m_1 - m_2}{1 + m_1 m_2} = \frac{-\frac{2}{14} - \frac{3}{2}}{1 + \left(-\frac{9}{14}\right)\frac{3}{2}} = \frac{-60}{1}$$

$$\theta = \tan^{-1}(60)$$

Example 136:

Show that the perpendicular from the origin upon the straight line joining the points (a cos α, a sin α) and (a cos β, a sin β) bisects the distance between them.

Solution:

The equation of the straight line joining the points (a cos α, a sin α) and (a cos β, a sin β) will be

$$y - a\sin\alpha = \frac{a\sin\beta - a\sin\alpha}{a\cos\beta - a\cos\alpha}(x - a\cos\alpha).$$

$$\text{or} \quad x\cos\frac{1}{2}(\alpha+\beta) + y\sin\frac{1}{2}(\alpha+\beta) = a\cos\frac{1}{2}(\alpha-\beta) \quad ...(1)$$

$$\text{Slope of (1)} = \frac{\cos\frac{1}{2}(\alpha+\beta)}{\sin\frac{1}{2}(\alpha+\beta)}$$

So slope of the line perpendicular to it will be + $\dfrac{\sin\frac{1}{2}(\alpha+\beta)}{\cos\frac{1}{2}(\alpha+\beta)}$

Hence the equation of the line passing through (0, 0) and perpendicular to (1) will be

$$y - 0 = \frac{\sin\frac{1}{2}(\alpha+\beta)}{\cos\frac{1}{2}(\alpha+\beta)}(x - 0)$$

$$\text{or} \quad x\sin\frac{1}{2}(\alpha+\beta) - y\cos\frac{1}{2}(\alpha+\beta) = 0 \quad ...(2)$$

To find out the point of intersection (1) and (2), we solve them; so putting the value of y from (2) in (1), we get

$$x\cos\frac{1}{2}(\alpha+\beta) + x\,.\frac{\sin^2\frac{1}{2}(\alpha+\beta)}{\cos\frac{1}{2}(\alpha+\beta)} = a\cos\frac{1}{2}(\alpha+\beta)$$

or $x\left[\cos^2\frac{1}{2}(\alpha+\beta)+\sin^2\frac{1}{2}(\alpha+\beta)\right]=a\cos\frac{1}{2}(\alpha+\beta).\cos\frac{1}{2}(\alpha+\beta)$

or $x=\frac{a}{2}\left[2\cos\frac{1}{2}(\alpha-\beta).\cos\frac{1}{2}(\alpha+\beta)\right]=\frac{a}{2}.\cos(\cos\alpha+\cos\beta)$

Similarly putting the value of x from (2) in (1), we get

$$y\left[\cos^2\frac{1}{2}(\alpha+\beta).\sin^2\frac{1}{2}(\alpha+\beta)\right]=a\cos\frac{1}{2}(\alpha-\beta).\sin\frac{1}{2}(\alpha+\beta)$$

or $y\,\frac{a}{2}\left[2\cos\frac{1}{2}(\alpha-\beta).\sin\frac{1}{2}(\alpha+\beta)\right]=\frac{a}{2}.(\sin\alpha+\sin\beta).$

Hence the point of intersection of (1) and (2) is

$$\left[\frac{a}{2}.(\cos\alpha+\cos\beta).\frac{a}{2}.(\sin\alpha+\sin\beta)\right]. \qquad ...(3)$$

Co-ordinates of the mid-point joining (a cos α, a sin α) and (a cos β, a sin β) will be

$$\left[\frac{1}{2}(a\cos\alpha+a\cos\beta),\frac{1}{2}(a\sin\beta+a\sin\beta)\right]$$

or $$\left[\frac{a}{2}(\cos\alpha+\cos\beta),\frac{a}{2}(\sin\alpha+\sin\beta)\right],$$

which are the same as given in (3). **Hence proved.**

Example 137:

Find the point of intersection and the inclination of the two lines

$$Ax + By = A + B$$

and $$A(x-y) + B(x+y) = 2B.$$

Solution:

The lines are

$$Ax + By = A + B$$

or $Ax + By - (A + B) = 0$ ...(1)

and $A(x-y) + B(x+y) = 2B$

or $x(A + B) + y(B - A) - 2B = 0$ ...(2)

By cross-multiplication, we get

$$\frac{x}{(B)(-2B)-(-A-B)(B-A)}=\frac{y}{(-A-B)(A+B)-(A)(-2B)}$$
$$=\frac{1}{(A)(B-A)-(B)(A+B)}$$

or $$\frac{x}{-\left(A^2+B^2\right)}=\frac{y}{-\left(A^2+B^2\right)}=\frac{1}{-\left(A^2+B^2\right)}$$

or $x = 1, y = 1$. So point of intersection is (1, 1) **Ans.**

For angle between them :

Slope of first line $= -\frac{A}{B} = m_1$ (say).

If θ be the angle between them, then

$$\tan\theta = \frac{-\frac{A}{B}+\frac{A+B}{B-A}}{1+\left(-\frac{A}{B}\right)\left(-\frac{A+B}{B-A}\right)} \quad \because \tan\theta = \frac{m_1-m_2}{1+m_1 m_2}$$

$$= \frac{-AB+A^2+AB+B^2}{B^2-AB+A^2+AB}$$

So $\theta = 45°$ **Ans.**

Example 138:

Find the distance of the point of intersection of the two straight lines $2x - 3y + 5 = 0$ and $3x + 4y = 0$ from the straight line $5x - 2y = 0$.

Solution:

The given straight lines are

$$2x - 3y + 5 = 0 \quad ...(1)$$

$$2x + 4y = 0 \quad ...(2)$$

and $$5x - 2y = 0. \quad ...(3)$$

Point of intersection of (1) and (2) is $\left(-\frac{20}{17}, \frac{15}{17}\right)$, {solving (1) and (2)}

To find out the distance of this point from (3), we substitute the co-ordinates $\left(-\frac{20}{17}, \frac{15}{17}\right)$ for x and y in (3) and divide by $\sqrt{[(5)^2 +(2)^2]}$. Hence the required distance is,

$$= \frac{5 \times \left(-\frac{20}{17}\right) - 2\left(\frac{15}{17}\right)}{\sqrt{\{(5)^2 + (2)^2\}}} = \frac{130}{17\sqrt{29}} \text{ (omitting –ve sing).}$$

Example 139:

Find the co-ordinates of the orthocentre of the triangle whose angular points are (0, 0), (2, –1) and (–1, 3).

Solution:

Let the points (0, 0), (2, –1) and (–1, 3) be A, B and C respectively. Let AL and BM be the perpendiculars from A and B on BC and CA produced respectively. Let these perpendiculars meet in Z.

Equation of AC is $y - 0 = \frac{3-0}{-1-0}(x - 0)$

or $y = -3x.$ ...(1)

Equation of BC is $y + 1 = \frac{3+1}{-1-2}(x - 2)$

or $(y + 1) = -\frac{4}{3}(x - 2)$ ...(2)

Any line passing through B, i.e. (2, –1) is $y + 1 = m(x - 2)$.

Slope of Ac is –3 [by (1)], so slope of BM is $\frac{1}{3}$ $(\because BM \perp AC)$

Hence equation of BM is $y + 1 = \frac{1}{3}(x - 2)$

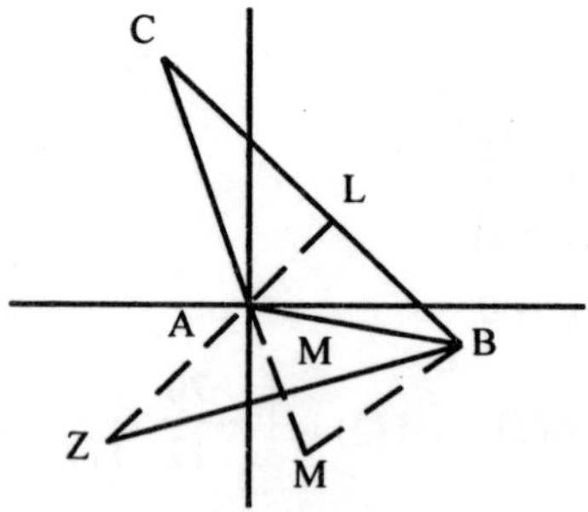

or $x - 3y = 5.$ ...(3)

Again, any line passing through (0, 0) is $y = mx$.

Slope of BC is $-\frac{4}{3}$ [by (2)]; so slope of AL is $\frac{3}{4}$ $(\because AL \perp BC)$

$\therefore$ The equation of AL is $y = \frac{3}{4}x$

or $3x - 4y = 0.$

Solving (3) and (4), we get $x = -4$, $y = -3$.

Hence the co-ordinates of the ortho-centre Z are (–4, –3). **Ans.**

Example 140:

Find the co-ordinates of the point of intersection of the straight lines $2x - 3y = 1$ and $5y - x = 3$, and determine also the angle at which they cut one another.

Solution:

The equations of the lines are

$$2x - 3y = 1, \qquad ...(1)$$

and

$$5y - x = 3, \qquad ...(2)$$

Multiplying (2) by 2 and adding to (1), we get $7y = 7$ or $y = 1$. Putting this value in (1), we get $x = 2$.

Hence the point of intersection of (1) and (2) is (2, 1).

Again slope of (1) is $\frac{2}{3}$ and slope of (2) is $\frac{1}{5}$.

If θ be the angle between them; then

$$\tan\theta = \frac{\frac{2}{3} - \frac{1}{5}}{1 + \frac{2}{3}\frac{1}{5}} = \frac{7}{17} \text{ or } \theta = \tan^{-1}\left(\frac{7}{17}\right)$$ **Ans.**

Example 142:

Prove that the following sets of three lines meet in a point $2x - 3y = 7$, $3x - 4y = 13$, and $8x - 11y = 33$.

Solution:

The equations of the lines are

$$2x - 3y = 7, \qquad ...(1)$$

$$3x - 4y = 13, \qquad ...(2)$$

and

$$8x - 11y = 33, \qquad ...(3)$$

Solving (1) and (2) simultaneously, we get $x = 11$, $y = 5$.

Substituting these values in (3), we get

L.H.S. = 8 × 11 – 11 × 5 = 33 = R.H.S.

Hence the three lines meet in a point.

Example 142:

Prove that the following sets of three lines meet in a point $3x - 4y + 6 = 0$, $6x - 5y + 9 = 0$, and $3x - 3y + 5 = 0$.

Solution:

The equations of the lines are

$$3x + 4y + 6 = 0, \quad ...(1)$$

$$6x + 5y + 9 = 0, \quad ...(2)$$

and $$3x + 3y + 5 = 0. \quad ...(3)$$

Solving (1) and (3) simultaneously, we get, $x = -\frac{2}{3}$, $y = -1$

Substituting these values in (2), we get

$$\text{L.H.S.} = 6 \times \left(-\frac{2}{3}\right) + 5 \times (-1) + 9 = 0 = \text{R.H.S.}$$

The equation is satisfied. **Hence Proved.**

Example 143:

Find the angle between the two lines $3x + y + 12 = 0$ and $x + 2y - 1 = 0$. Find also the co-ordinates of their point of intersection and the equations of lines drawn perpendicular to them from the point (3, –2).

Solution:

The two lines are

$$3x + y + 12 = 0 \quad ...(1)$$

and $$x + 2y - 1 = 0 \quad ...(2)$$

Their slopes are $\left(-\frac{3}{1}\right)$ and $\left(-\frac{1}{2}\right)$ respectively. So if q be the angle between them,

$$\tan\theta = \frac{\left(-\frac{3}{1}\right) + \left(-\frac{1}{2}\right)}{1 + \left(-\frac{3}{1}\right)\left(-\frac{1}{2}\right)} = -1.$$

So $\theta = 135°$, or the cute angle between them will be 45°. **Ans.**

Multiplying (2) by 3 and subtracting from (1), we get $y = 3$, and putting this value in (1), we get $x = -5$.

Hence the point of intersection of (1) and (2) is (–5, 3). **Ans.**

Again, any line passing through (3, –2) will be $y + 2 = m(x - 3)$ ...(3)

Slope of (1) is $-3/1$ and if (3) is perpendicular to (1), then

$$m \times -3 = -1 \text{ or } m = \tfrac{3}{1}.$$

Substituting this value in (3), we get $y + 2 = 1/3\ (x - 3)$.

Hence the line passing through (3), (2) and perpendicular to (1) is $x - 3$.

Again slope of (2) is $(-1/2)$ and if (3) is perpendicular to (2), then

$$m \times -\ 1/2 = -1$$

or $\qquad m = 2.$

Putting this value in (3), we get $y + 2 = 2\ (x - 3)$.

So the equation of the line passing through $(3, -2)$ and perpendicular to (2) is $2x - y = 8$, **Ans.**

Example 144:

Find the co-ordinates of the feet of the perpendiculars let fall from the point (5, 0) upon the sides of the triangle formed by joining the three points (4, 3), (–4, 3), and (0, –5); prove also that the points so determined lie on a straight line.

Solution:

Let the point P be (5, 0) and (4, 3), (–4, 3) and (0, –5) be A, B and C respectively.

Equation of the line AB will be

$$y - 3 = \frac{3 - 3}{4 + 4}(x - 4)$$

or $\qquad y - 3 = 0. \qquad \qquad ...(1)$

Equation of the line BC will be

$$y - 3 = \frac{-5 - 3}{0 + 4}(x + 4).$$

or $\qquad 2x + y + 5 = 0. \qquad \qquad ...(2)$

Equation of the line CA is $y + 5 = \dfrac{3 + 5}{4 - 0}(x - 0)$

or $\qquad 2x - y - 5 - 0. \qquad \qquad ...(3)$

Equation of any line passing through P (5, 0) is

$$y - 0 = m\ (x - 5) \qquad \qquad ...(4)$$

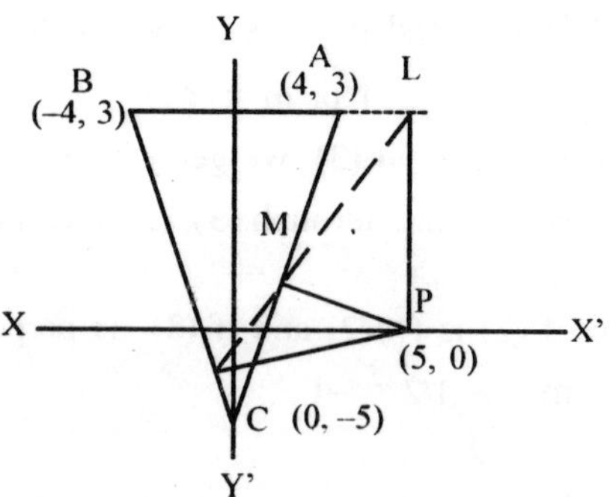

Side AB [given by (1)] is parallel to axis of x; hence PL the line perpendicular to it, will be parallel to axis of y, and as it passes through (5, 0), its equation will be

$$x - 5 = 0 \quad \text{or } x = 5. \qquad ...(5)$$

Solving (1) and (5), we get the co-ordinates of L as (5, 3).

Again, slope of BC [given by (2)] is –2.

Hence slope of PN = $\frac{1}{2}$ (where PN is perpendicular to BC).

Putting this value in (4), we get the equation of PL.

$$y = \frac{1}{2}(x - 5) \quad \text{or} \quad x - 2y - 5 = 0. \qquad ...(6)$$

Solving (2) and (6), the co-ordinates of N are (–1, –3).

Again the slope of AC [given by (3)] is $\frac{2}{1}$.

Hence slope of PM will be $\left(-\frac{2}{1}\right)$ (where PM is perpendicular to AC).

Putting this value in (4), we get the equation of PM as

$$y = \frac{1}{2}(x - 5) \text{ or } x + 2y - 5 = 0. \qquad ...(7)$$

Solving (3) and (7), we get the co-ordinates of M as (3, 1).

Hence the co-ordinates of L, N and M are (5, 3), (–1, –3) and (3, 1) respectively.

Again, the equation of the line passing through L and M is

$$y - 5 = \frac{1 - 3}{3 - 5}(x - 5) \quad \text{or} \quad x - y - 2 = 0 \qquad ...(8)$$

Putting the co-ordinates of M (–1, –3) in L.H.S. of (8), we find

$$-1 - (-3) - 2 = 0 \qquad ...(9)$$

As the co-ordinates satisfy, hence M lies on the straight line LN or L, M and N are collinear. **Proved.**

Example 145:

Prove that the following sets of three lines meet in a point

$$\frac{x}{a} + \frac{y}{b} = 1, \frac{x}{b} + \frac{y}{a} = 1, \text{ and } y = x.$$

Solution:

The equations of the lines are

$$\frac{x}{a} + \frac{y}{b} = 1 \text{ or } bx + ay - ab = 0 \qquad ...(1)$$

$$\frac{x}{b} + \frac{y}{a} = 1 \text{ or } ax + by - ab = 0 \qquad ...(2)$$

and $$x - y = 0 \qquad ...(3)$$

Solving (2) and (3) simultaneously, we get $\left(x = \frac{ab}{a+b}, y = \frac{ab}{a+b}\right)$

Substituting in (1), we get the L.H.S. as

$$= b = \frac{ab}{a+b} + a \cdot \frac{ab}{a+b} - ab = \frac{ab}{a+b}(b + a) - ab = 0 = \text{R.H.S.}$$

So the equation is satisfied. **Hence proved.**

Example 146:

Find the conditions that the straight lines $y = m_1x + a_1$, $y = m_2x + a_2$ and $y = m_3x + a_3$ may meet in a point.

Solution:

The equations of the given straight lines are

$$y = m_1x + a_1 \qquad ...(1)$$

$$y = m_2x + a_2 \qquad ...(2)$$

and $$y = m_3x + a_3 \qquad ...(3)$$

respectively.

To find out the point of intersection of (1) and (2), we solve them simultaneously. Solving, we get $x = -\frac{a_1 - a_2}{m_1 - m_2}$ and $y = \frac{a_2m_1 - a_1m_2}{m_1 - m_2}$.

If the three lines meet in a point, these values of x and y (i.e., the co-ordinates of point of intersection) will satisfy the third. Hence substituting in (3), we get

$$\frac{a_2m_1 - a_1m_2}{m_1 - m_2} = m_3\left(-\frac{a_1 - a_2}{m_1 - m_2}\right) + a_3$$

or $$a_2m_1 - a_1m_2 = -a_1m_3 + a_2m_1 - a_3m_2$$

or $$m_1(a_2 - a_3) + m_2(a_3 - a_1) + m_3(a_1 - a_2) = 0.$$ **Ans.**

Example 147:

Prove that the three straight lines whose equations are 15x – 18y + 1 = 0, 12x + 10y – 3 = 0, and 6x + 66y – 11 = 0 all meet in a point.

Show also that the third line bisects the angle between the other two.

Solution:

The equations are

$$15x - 18y + 1 = 0 \quad ...(1)$$

$$12x + 10y - 3 = 0 \quad ...(2)$$

$$6x + 66y - 11 = 0 \quad ...(3)$$

Solving (2) and (3) simultaneously, we get $x = \frac{22}{183}$ and $y = \frac{19}{122}$.

Substituting in (1), we get the L.H.S. as $15 \times \frac{22}{183} - 18 \times \frac{19}{122} + 1 = 0$

or $$\frac{110}{61} - \frac{171}{61} + 1 = 0 = \text{R.H.S.}$$

So the equation is satisfied. **Hence proved.**

The equation is satisfied.

The equation of the bisector of (1) and (2) is

$$\frac{15x - 18y + 1}{\sqrt{\left[(15)^2 + (18)^2\right]}} = \pm\frac{12x + 10y - 3}{\sqrt{\left[(12)^2 + (10)^2\right]}}$$

or $$\frac{15x - 18y + 1}{3\sqrt{61}} = \pm\frac{12x + 10y - 3}{2\sqrt{61}}$$

or $$2(15x - 18y + 1) = 3(12x + 10y - 3)$$ (taking +ve sing) or $\qquad 6x + 66y - 11 = 0$ which is same as (3).

Example 148:

Find the co-ordinates of the orthocentre of the triangle whose angular points are (1, 0), (2, –4) and (–5, –2).

Solution:

Let the given points (1, 0), (2, –4) and (–5, –2) be A, B and C respectively. Let AL and BM be the perpendiculars from A and B on BC and AC respectively and AL and BM meet in Z.

The equation of AC will be

$$y = 0 = \frac{-2-0}{-5-1}(x-1)$$

or $\qquad x - 3y - 1 = 0. \qquad$ ...(1)

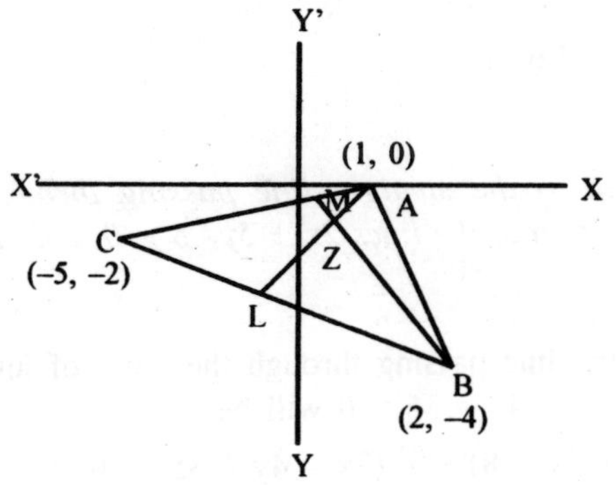

Hence slope of AC is $\frac{1}{3}$.

Equation of any line passing through B (2, –4) is $y + 4 = m(x - 2)$.

As BM is perpendicular to AC, the slope of BM will be –3.

So equation of BM is $y + 4 = -3(x - 2)$

or $\qquad 3x + y = 2. \qquad$ ...(2)

Again slope of the line joining B and C is $\dfrac{-2+4}{-5-2} = -\dfrac{2}{7}$.

So slope of AL $= \frac{7}{2}$ (as AL is perpendicular to BC)

Equation of anytime passing through A is $y - 0 = m(x - 1)$

Substituting m for AL we have $y = \frac{7}{2}(x - 1)$

or $7x - 2y = 7$ ...(3)

Solving (2) and (3), we get $x = \frac{11}{13}$, $y = -\frac{7}{13}$.

Therefore, the co-ordinates of the ortho-centre Z are $\left(\frac{11}{13}, -\frac{7}{13}\right)$. **Ans.**

Example 149(a):

Find the equation to the straight line passing through the point (3, 2) and the point of intersection of the lines 2x + 3y = 1 and 3x – 4y = 6.

Solution:

Equation of any line passing through the point of intersection of $2x + 3y - 1 = 0$ and $3x - 4y - 6 = 0$ will be

$$(2x + 3y - 1) + \lambda\,(3x - 4y - 6) = 0. \qquad ...(1)$$

As the line passes through (3, 2), the co-ordinates will satisfy (1), hence

$$(2.3 + 3.2 - 1) + 1\,(3.3 - 4.2 - 6) = 0 \quad \text{or } \lambda = \frac{11}{5}.$$

Substituting in (1), we get, $(2x + 3y - 1) + \frac{11}{5}(3x - 4y - 6) = 0$

or $43x - 29y = 71.$ **Ans.**

Example 149(b):

Find the equation to the straight line passing through the point (2, –9) and the intersection of the lines 2x + 5y– 8 = 0 and 3x – 4y = 35.

Solution:

The equation of the line passing through the point of intersection of $2x + 5y - 8 = 0$ and $3x - 4y - 35 = 0$ will be

$$(2x + 5y - 8) + \lambda\,(3x - 4y - 35) = 0 \qquad ...(1)$$

As the line passes through (2, –9), the co-ordinates will satisfy (1); hence

$$[(2 \times 2) + (5 \times -9) - 8] + \lambda\,[(3 \times 2) - (4 \times -9) - 35] = 0$$

or $\lambda = 7.$

Substituting in (1), we get

$$(2x + 5y - 8) + 7\,(3x - 4y - 35) = 0$$

or $23x - 23y - 253 = 0$

or $x - y = 11.$ **Ans.**

Example 150:

Find the equation to the straight line passing through the intersection of the lines x + 2y + 3 = 0 and 3x + 4y + 7 = 0 and perpendicular to the straight line y – x = 8.

Solution:

The equation of the line passing through the point of intersection of $x + 2y + 3 = 0$ and $3x + 4y + 7 = 0$ will be

$$(x + 2y + 3) + \lambda\,(3x + 4y + 7) = 0$$

or $$x\,(1 + 3\lambda) + y\,(2 + 4\lambda) + 3 + 7\lambda = 0. \quad ...(1)$$

Slope of (1) $$m_1 = -\frac{1+3\lambda}{2+4\lambda}. \quad ...(2)$$

The given line is $\quad y - x = 8 \quad ...(3)$

Slope of given line $\quad m_2 = +1. \quad ...(4)$

As (1) is perpendicular to (3), we have $m_1 \times m_2 = -1$

or $$-\frac{1+3\lambda}{2+4\lambda} \times 1 = -1 \quad \text{or } \lambda = -1.$$

Substituting in (1), we get

$$x\,(1 - 3) + y\,(2 - 4) + 3 - 7 \quad \text{or} \quad x + y + 2 = 0.$$ **Ans.**

Example 151:

In any triangle ABC, prove that

(i) *the bisectors of the angles A, B and C meet in a point,*

(ii) *the medians, i.e. the lines joining each vertex to the middle point of the opposite side, meet in a point, and*

(iii) *the straight lines through the middle points of the sides perpendicular to the sides meet in a point.*

Solution:

(i) Let the triangle be ABC and the equations of the sides AB, BC and CA be

$$a_1x + b_1x + c_1 = 0 \quad ...(1)$$

$$a_2x + b_2x + c_2 = 0 \quad ...(2)$$

and $$a_3x + b_3x + c_3 = 0 \quad ...(3)$$

respectively.

Equation of the bisector of $\angle ABC$

$$= \frac{a_1x + b_1x + c_1}{\sqrt{(a_1^2 + b_1^2)}} = \pm\frac{a_2x + b_2x + c_2}{\sqrt{(a_2^2 + b_2^2)}} \quad \text{[bisector of (1) and (2)]}$$

or $$\frac{a_1x + b_1x + c_1}{\sqrt{\left(a_1^2 + b_1^2\right)}} - \frac{a_2x + b_2x + c_2}{\sqrt{\left(a_2^2 + b_2^2\right)}} = 0 \text{ (taking +ve sign)}$$

or $$x\left(\frac{a_1}{\sqrt{\left(a_1^2 + b_1^2\right)}} - \frac{a_2}{\sqrt{\left(a_2^2 + b_2^2\right)}}\right) + y\left(\frac{b_1}{\sqrt{\left(a_1^2 + b_1^2\right)}} - \frac{b_2}{\sqrt{\left(a_2^2 + b_2^2\right)}}\right)$$

$$+\left(\frac{c_1}{\sqrt{\left(a_1^2 + b_1^2\right)}} - \frac{c_2}{\sqrt{\left(a_2^2 + b_2^2\right)}}\right) = 0 \quad ...(1)$$

Similarly the bisector of ∠BCA, i.e. equation (2) and (3) is

$$x\left(\frac{a_2}{\sqrt{\left(a_2^2 + b_2^2\right)}} - \frac{a_3}{\sqrt{\left(a_3^2 + b_3^2\right)}}\right) + y\left(\frac{b_2}{\sqrt{\left(a_2^2 + b_2^2\right)}} - \frac{b_3}{\sqrt{\left(c_3^2 + b_3^2\right)}}\right)$$

$$+\left(\frac{c_2}{\sqrt{\left(a_2^2 + b_2^2\right)}} - \frac{c_3}{\sqrt{\left(a_3^2 + b_3^2\right)}}\right) = 0 \quad ...(2)$$

and the bisector of ∠CAB, i.e. (3) and (1) is

$$x\left(\frac{a_3}{\sqrt{\left(a_3^2 + b_3^2\right)}} - \frac{a_1}{\sqrt{\left(a_1^2 + b_1^2\right)}}\right) + y\left(\frac{b_3}{\sqrt{\left(a_3^2 + b_3^2\right)}} - \frac{b_1}{\sqrt{\left(a_1^2 + b_1^2\right)}}\right)$$

$$+\left(\frac{c_3}{\sqrt{\left(c_3^2 + b_3^2\right)}} - \frac{c_1}{\sqrt{\left(a_1^2 + b_1^2\right)}}\right) = 0. \quad ...(3)$$

Adding (1), (2) and (3), we find that the sums of coefficients of x and y and the constant term are respectively 0. So, it becomes identically 0.

Hence the three lines given by (1), (2) and (3) meet in a point.

Proved.

(ii) Let the points A, B and C be (x_1, y_1), (x_2, y_2) and (x_3, y_3) respectively.

If D, E, F be the mid-points of BC, CA and AB respectively, then co-ordinates of D are $\left(\frac{x_2 + x_3}{2}, \frac{y_2 + y_3}{2}\right)$

Similarly, co-ordinates of E and F are

$$\left(\frac{x_1+x_3}{2}, \frac{y_1+y_3}{2}\right) \text{ and } \left(\frac{x_1+x_2}{2}, \frac{y_1+y_2}{2}\right) \text{ respectively.}$$

Equation of AD will be $y - y_1 = \dfrac{\dfrac{y_2+y_3}{2} - y_1}{\dfrac{x_2+x_3}{2} - x_1}(x - x_1)$

or $$y - y_1 = \frac{y_2 + y_3 - 2y_1}{x_2 + x_3 - 2x_1}(x - x_1)$$

or $$(y - y_1)(x_2 + x_3 - 2x_1) = (y_2 + y_3 - 2y_1)(x - x_1)$$

or $$x(y_2 + y_3 - 2y_1) - y(x_1 + x_3 - 2x_1)$$
$$x_2y_1 - x_1y_2 + y_1x_3 - x_1y_3 = 0 \qquad ...(1)$$

Similarly, the equations of BE and CF respectively are

$$x(y_3 + y_1 - 2y_2) - y(x_3 + x_1 - 2x_2)$$
$$+ x_3y_2 - x_2y_3 + y_2x_1 - x_2y_1 = 0 \qquad ...(2)$$

and $$x(y_1 + y_2 - 2y_3) + y(x_1 + y_2 - 2x_3)$$
$$+ x_1y_3 - x_3y_1 + y_3x_2 - x_3y_2 = 0 \qquad ...(3)$$

Adding (1), (2) and (3), we get an equation which is identically equal to zero; hence AD, BE and CF meet in a point. **Proved.**

(iii) Let ABC be the triangle and the co-ordinates of A, B and C be (x_1, y_1), (x_2, y_2) and (x_3, y_3) respectively. Co-ordinates of D, E, F the mid points of BC, CA and AB will be $\left(\dfrac{x_2+x_3}{2}, \dfrac{y_2+y_3}{2}\right)$; $\left(\dfrac{x_3+x_1}{2}, \dfrac{y_3+y_1}{2}\right)$ and $\left(\dfrac{x_1+x_2}{2}, \dfrac{y_1+y_2}{2}\right)$ respectively.

Slope of the line BC $= \dfrac{y_3+y_2}{x_3-x_2}$.

$\therefore$ Slope of line perpendicular to BC will be $-\dfrac{x_3+x_2}{y_3-y_2}$.

Therefore, the equation of the line passing through D and perpendicular to BC will be

$$y - \frac{y_2 + y_3}{2} = -\frac{x_3 + x_2}{y_3 - y_2}\left(x - \frac{x_2 + x_3}{2}\right).$$

or $$y\,(y_3 - y_2) + x\,(x_3 - x_2) - \frac{1}{2}\left(y_2^2 - y_3^2 + x_2^2 - x_3^2\right) = 0. \qquad ...(1)$$

Similarly, the equation of the line perpendicular to AC passing through E will be

$$y\,(y_1 - y_3) + x\,(x_1 - x_3) - \frac{1}{2}\left(y_3^2 - y_1^2 + x_3^2 - x_1^2\right) = 0. \qquad ...(2)$$

and the equation of the line perpendicular to AB and passing through F will be

$$y\,(y_2 - y_1) + x\,(x_2 - x_1) - \frac{1}{2}\left(y_1^2 - y_2^2 + x_1^2 - x_2^2\right) = 0. \qquad ..(3)$$

Adding (1), (2) and (3), we find an equation which is identically zero; hence the 3 lines meet in a point. **Hence proved.**

Example 152:

Find the equation to the straight line passing through the origin and the point of intersection of $x - y - 4 = 0$ and $7x + y + 20 = 0$, proving that it bisects the angle between them.

Solution:

The equation of the line passing through the point of intersection of

$$x - y - 4 = 0 \qquad ...(1)$$

and $$7x + y + 20 = 0 \qquad ...(2)$$

will be $$(x - y - 4) + \lambda\,(7x + y + 20) = 0. \qquad ...(3)$$

As the line passes through origin, hence (0, 0) will satisfy it; hence substituting the co-ordinates in (3), we get

$$(0 - 0 - 4) + \lambda\,(7 \times 0 + 0 + 20) = 0 \text{ or } \lambda = \frac{1}{5}.$$

Substituting in (3), we get $(x - y - 4) + \frac{1}{5}(7x + y + 2) = 0$

or $$12x - 4y = 0 \text{ or } y = 3x. \qquad ...(4)$$

Equation of the line bisecting the angle between (1) and (2) will be

$$\frac{x - y - 4}{\sqrt{(1 + 1)}} = -\frac{7x + y + 20}{\sqrt{(49 + 1)}}.$$

or $$\frac{x - y - 4}{\sqrt{(2)}} = -\frac{7x + y + 20}{5\sqrt{2}} \qquad \text{(taking –ve sign only)}$$

or $12x - 4y = 0$

or $y = 3x$, which is same as (4). **Hence proved.**

Example 153:

Prove the same property for the parallelogram whose sides are

$$\frac{x}{a} + \frac{y}{b} = 1, \frac{x}{b} + \frac{y}{a} = 1, \frac{x}{a} + \frac{y}{b} = 2 \text{ and } \frac{x}{b} + \frac{y}{a} = 2.$$

Solution:

Let the given lines

$$\frac{x}{a} + \frac{y}{b} = 1, \quad ...(1)$$

$$\frac{x}{b} + \frac{y}{a} = 1, \quad ...(2)$$

$$\frac{x}{a} + \frac{y}{b} = 2, \quad ...(3)$$

$$\text{and } \frac{x}{b} + \frac{y}{a} = 2 \quad ...(4)$$

represent AB, BC, CD and DA respectively.

Solving (4) and (1), we get the co-ordinates of A as

$$\left\{\frac{ab\,(2a-b)}{a^2 - b^2}, \frac{ab(a - 2b)}{a^2 - b^2}\right\}$$

Solving (1) and (2), we get the co-ordinates of B as $\left\{\frac{ab}{a + b}, \frac{ab}{a + b}\right\}$.

Solving (2) and (3), we get the co-ordinates of C as

$$\left\{\frac{ab\,(a-2b)}{a^2 - b^2}, \frac{ab(2a - b)}{a^2 - b^2}\right\}.$$

Solving (3) and (4), we get the co-ordinates of D as $\left\{\frac{2ab}{a + b}, \frac{2ab}{a + b}\right\}$.

$$\text{Slope of the diagonal AC} = m_1 = \frac{\frac{ab(2a - b)}{a^2 - b^2} - \frac{ab(a - 2b)}{a^2 - b^2}}{\frac{ab(a - 2b)}{a^2 - b^2} - \frac{ab(2a - b)}{a^2 - b^2}}$$

$$\left\{\because m = \frac{y_2 - y_1}{x_2 - x_1}\right\}$$

or $$m_1 = \frac{ab(2a - b - a + 2b)/(a^2 - b^2)}{ab\,(a - 2b - 2a + b)/(a^2 - b^2)} = -1. \qquad ...(5)$$

Slope of the diagonal BD = $m_2 = \dfrac{\{2ab/(a + b)\} - \{ab/(a + b)\}}{\{2ab/(a + b)\} - \{ab/(a + b)\}} = 1.$

...(6)

Since $m_2 \times m_1 = -1$, hence AC and BD are perpendicular to each other.

Example 154:

Find the equation to the straight line passing through the intersection of the lines $x - 2y - a = 0$ and $x + 3y - 2a = 0$, and parallel to the straight line $3x + 4y = 0$.

Solution:

The equation of the line passing through the point of intersection of $x - 2y - a = 0$ and $x + 3y - 2a = 0$ will be

$$(x - 2y - a) + 1\,(x + 3y - 2a) = 0$$

or $$x\,(1 + \lambda) + y\,(3\lambda - 2) - a - 2a\lambda = 0 \qquad ...(1)$$

Slope of (1) = $-\dfrac{1 + \lambda}{3\lambda - 2} = \dfrac{1 + \lambda}{2 - 3\lambda}$

The given equation is $3x + 4y = 0$ ...(2)

Slope of (2) = $-\frac{3}{4}$.

As (1) is parallel to (2), so slopes are same.

$$\therefore \quad \frac{1 + \lambda}{2 - 3\lambda} = -\frac{3}{4} \text{ or } \lambda = 2.$$

Substituting in (1), we get $x\,(1 + 2) + y\,(6 - 2) - a - 4a = 0$

or $3x + 4y = 5a.$ **Ans.**

Example 155:

Find the equation to the straight line passing through the intersection of the lines $3x - 4y + 1 = 0$ and $5x + y - 1 = 0$ and cutting off equal intercepts from the axes.

Solution:

The given lines are

$$3x - 4y + 1 = 0 \quad ...(1)$$

and $$5x + y - 1 = 0. \quad ...(2)$$

Solving (1) and (2), we get $x = \frac{3}{23}, y = \frac{8}{23}$.

Hence the point of intersection of (1) and (2) is $\left(\frac{3}{23}, \frac{8}{23}\right)$.

As the line cuts off equal intercepts from the axes, let each intercept be a.

Let the equation of the straight line be

$$\frac{x}{a} + \frac{y}{a} = 1 \quad \text{or} \quad x + y = a \quad ...(3)$$

It passes through $\left(\frac{3}{23}, \frac{8}{23}\right)$

$$\therefore \quad \frac{3}{23} + \frac{8}{23} = a \quad \text{or} \quad a = \frac{11}{23}$$

Substituting the value of a in (3) the equation of line is

$$x + y = \frac{11}{23} \quad \text{or} \quad 23x + 23y = 11.$$

Example 156(a):

Prove that the diagonals of the parallelogram formed by the parallelogram formed by the four straight lines $\sqrt{3}x + y = 0$, $\sqrt{3}y + x = 0$, $\sqrt{3}x + y = 1$, and $\sqrt{3}y + x = 1$, are at right angles to one another.

Solution:

Let the given lines

$$\sqrt{3}x + y = 0, \quad ...(1)$$

$$\sqrt{3}Y + x = 0, \quad ...(2)$$

$$\sqrt{3}x + y = 1, \quad ...(3)$$

$$\sqrt{3}y + x = 1, \quad ...(4)$$

and

be AB, BC, CD and DA respectively.

Solving (1) and (2), we get the co-ordinates of B as (0, 0)

Solving (2) and (3), we get the co-ordinates of C as $\left(\frac{\sqrt{3}}{2}, \frac{1}{2}\right)$.

Solving (3) and (4), we get the co-ordinates of D as $\left(\frac{\sqrt{3}-1}{2}, \frac{\sqrt{3}-1}{2}\right)$.

Ans.

Solving (4) and (1), we get the co-ordinates of A as $\left(-\frac{1}{2}, \frac{\sqrt{3}}{2}\right)$.

Slope of the diagonal AC = $m_1 = \dfrac{\frac{\sqrt{3}}{2} - \left(-\frac{1}{2}\right)}{\left(-\frac{1}{2}\right) - \frac{\sqrt{3}}{2}}$. $\left\{\because m = \dfrac{y_2 - y_1}{x_2 - x_1}\right\}$

$\therefore \qquad m1 = -1.$ ...(5)

Slope of the diagonal BD = $m_2 = \dfrac{\left\{\left(\sqrt{3}-1\right)/2\right\} - 0}{\left\{\left(\sqrt{3}-1\right)/2\right\} - 0}$

or $\qquad m_2 = 1$

Since $m_1 \times m_2 = -1$, hence AC and BD are perpendicular to each other.

Proved.

Example 156(b):

Find the equation to the straight line passing through the origin and the point of intersection of the lines $\frac{x}{a} + \frac{y}{b} = 1$ and $\frac{x}{b} + \frac{y}{a} = 1$.

Solution:

The two equations are

$$\frac{x}{a} + \frac{y}{b} = 1 \text{ or } bx + ay - ab = 0 \qquad ...(1)$$

and
$$\frac{x}{b} + \frac{y}{a} = 1 \text{ or } ax + by - ab = 0. \qquad ...(2)$$

The equation of a line passing through the point of intersection of (1) and (2) will be $(bx + ay - ab) + \lambda\,(ax + by - ab) = 0$. ...(3)

As the required line passes through origin, hence (0, 0) will satisfy (3); so $(0 + 0 - ab) + \lambda\,(0 + 0 - ab) = 0$ or $\lambda = -1$.

Substituting in (3), we get, $(bx + ay - ab) - 1\ (ax + by - ab) = 0$

or $\quad x\ (b - a) - y\ (b - a) = 0$ or $y = x.$ **Ans.**

Example 156(c):

Find the equation to the straight line passing through the point (a, b) an the intersection of the same two lines.

Solution:

The equation is same as (3) in Q. No. 142, i.e.

$$(bx + ay - ab)\ \lambda\ (ax + by - ab) = 0 \qquad ...(1)$$

As this straight line passes through (a, b), so the co-ordinates will satisfy (1); hence

$$(b.a + a.b - ab) + \lambda\ (a.a + b.b - ab) = 0$$

or $$\lambda = \frac{ab}{ab - a^2 - b^2}.$$

Substituting in (1), we get

$$(bx + ay - ab) + \frac{ab}{ab - a^2 - b^2}\ (ax + by - ab) = 0$$

or $$b\ (ab - a^2 - b^2)\ x + a\ (ab - a^2 - b^2)\ y - ab\ (ab - a^2 - b^2) + ab\ (ax + by - ab) = 0$$

or $$(ab^2 - b^3)\ x + (a^2b - a^3)y + a^3b + ab^3 - 2a2b2 = 0$$

or $$b^2\ (a - b)\ x - a^2\ (a - b)\ y + ab\ (a - b)^2 = 0$$

or $$a^2y - b^2x = ab\ (a - b).$$ **Ans.**

Example 156(d):

Find the equation to the straight line passing through the intersection of the lines 2x – 3y = 10 and x + 2y = 6 and the intersection of the lines 16x – 10y = 33 and 12x + 14y + 29 = 0.

Solution:

The first two equations are

$$2x - 3y = 10 \qquad ...(1)$$

and $$x + 2y = 6. \qquad ...(2)$$

Solving (1) and (2), we get $x = \frac{38}{7}, y = \frac{2}{7}.$

Hence the point of intersection of (1) and (2) is $\left(\frac{38}{7}, \frac{2}{7}\right).$

Again the other two straight lines are given as

$$16x - 10y = 33 \qquad ...(3)$$

and $$12x + 14y + 29 = 0 \qquad ...(4)$$

Solving (3) and (4), we get $x = \frac{1}{2}, y = -\frac{5}{2}$.

Hence the point of intersection of (3) and (4) is $\left(\frac{1}{2}, -\frac{5}{2}\right)$.

The equation f the line passing through the point of intersection of (1) and (2), and (3) and (4) will be

$$y - \frac{2}{7} = \frac{-\frac{5}{2} - \frac{2}{7}}{\frac{1}{2} - \frac{38}{7}}\left(x - \frac{38}{7}\right)$$

or $$(7y - 2) = \frac{-39}{-69}(7x - 38)$$

or $$13x - 23y = 64.$$ **Ans.**

Example 157:

Find the equations to the straight lines passing through the point of intersection of the straight lines Ax + By + C = 0 and A'x + B'y + C' = 0 and

(i) passing through the origin,

(ii) parallel to the axis of y,

(iii) cutting off a given distance a from the axis f y, and

(iv) passing through a given point (x', y').

Solution:

The equation of any line passing through the pint of intersection of Ax + By + C = 0 and A'x + B'y + C' = 0 will be

$$(Ax + By + C) + \lambda\,(A'x + B'y + C') = 0. \qquad ...(1)$$

(i) If the ling given by (1) passes through (0, 0), the co-ordinates will satisfy (1).

Hence $(0 + 0 + C) + \lambda\,(0 + 0 + C') = 0$ or $\lambda = -\frac{C}{C'}$.

Putting in (1), we get $(Ax + By + C) - \frac{C}{C'}(A'x + B'y + C') = 0$. **Ans.**

(ii) If the required line is parallel to axis of y, its slope will be ∞ (infinity).

Slope of (1) $= -\frac{A + A'\lambda}{B + B'\lambda} = \infty$; $\therefore \lambda = -\frac{B}{B'}$ (∵ Deno. = 0).

Substituting in (1), we get $(Ax + By + C) - \frac{B}{B'}(A'x + B'y + C') = 0.$

(iii) If the line given by (1) cuts off an intercept of a from y-axis, then the point f intersection with y-axis will be (0, a).

Substituting these co-ordinates in (1), we get

$$(Ba + C) + \lambda (B'a + C') = 0 \text{ or } \lambda = -\frac{Ba + C}{Ba' + C'}$$

Hence the required equation is

$$(Ax + By + C) - \frac{Ba + C}{Ba' + C'}(Ax' + By' + C') = 0. \text{ Ans.}$$

(iv) If the line given by (1) passes through (x', y'), the co-ordinates will satisfy (1). Hence substituting, we get

$$(Ax' + BY' + C') + \lambda (A'x' + B'y' + C') = 0$$

or $$\lambda = \frac{Ax' + By' + C}{A'x' + B'y' + C'}$$

Hence the required equations

$$Ax + By + C \frac{Ax' + By' + C}{A'x' + B'y' + C'}(A'x + B'y + C') = 0. \qquad \text{Ans.}$$

Example 158(a):

Find the equations to the straight lines bisecting the angles between the following pairs of straight lines, playing first the bisector of the angle in which the origin lines 12x + 5y – 4 = 0 and 3x + 4y + 7 = 0.

Solution:

The given equation are

$$12x + 5y - 4 = 0, \text{ and } 3x + 4y + 7 = 0$$

or $$-12x - 5y + 4 = 0 \qquad ...(1)$$

and $$3x + 4y + 7 = 0 \qquad ...(2)$$

The equation of the bisector of that angle in which the origin lines will be

$$\frac{-12x - 5y + 4}{\sqrt{\{(-12)^2 + (-5)^2\}}} = +\frac{3x + 4y + 7}{\{(3)^2 + (4)^2\}}$$

or $$\frac{-12x - 5y + 4}{13} = \frac{3x + 4y + 7}{5}$$

or $$99x + 77x + 71 = 0 \qquad ...(3)$$

The equation of the other bisector will be

$$\frac{-12x - 5y + 4}{13} = -\frac{3x + 4y + 7}{5}$$

or $$21x - 27y - 111 = 0 \text{ or } 7x - 9y - 37 = 0. \qquad ...(4)$$

Hence the required equations are

$$9x + 77y + 71 = 0$$

and $$7x - 9y - 37 = 0.$$ **Ans.**

Example 158(b):

If through the angular points of a triangle straight lines be drawn parallel to the sides, and if the intersections of these lines be joined to the opposite angular points of the triangle, show that the joining lines so obtained will meet in a point.

Solution:

Let ABC be the triangle such that base BC = 2a. Taking axis of x coinciding with BC and the line perpendicular to BC through the mid-point of BC as y-axis, suppose the co-ordinates of vertex A be (x_1, y_1). The co-ordinates of B and C will be (–a, 0) and (a, 0) respectively.

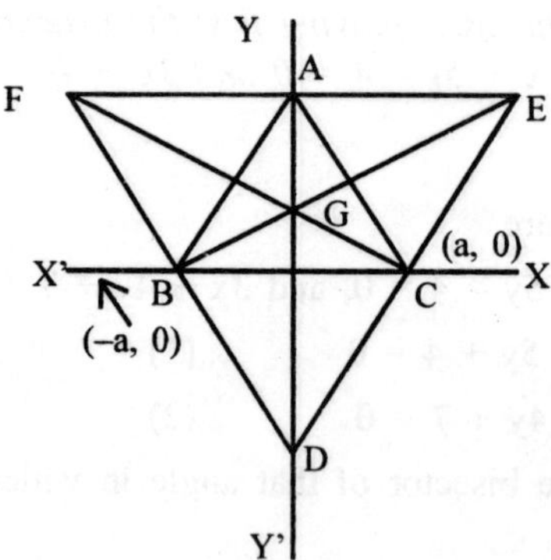

Let DE, EF and FD be the lines parallel to AB, BC and CA respectively and passing through C, A and B respectively.

Slope of AB $= \dfrac{y_1}{x_1 + a}$ ...(1)

Slope of BC $= 0$ ...(2)

and slope of CA $= \dfrac{y_1}{x_1 - a}$. ...(3)

Equation of EF is $y - y_1 = 0\,(x - x_1)$

or $y = y_1$ ...(4)

Again equation of FD will be

$$y - 0 = \frac{y_1}{x_1 - a}\,(x + a)$$

or $xy_1 - y\,(x_1 - a) + ay_1 = 0$. ...(5)

Again the equation of ED will be

$$y - 0 = \frac{y_1}{x_1 + a}\,(x + a)$$

passing through A $= (x_1, y_1)$ and parallel to BC ($m = 0$)

passing through B $= (-a, 0)$ and parallel to AC, $m = \dfrac{y_1}{x_1 + a}$

passing through B $= (a, 0)$ and parallel to AB, $m = \dfrac{y}{x_1 + a}$.

or $xy_1 - y\,(x_1 + a) - ay_1 = 0$. ...(6)

Solving (4) and (5), we get $x = (x_1 - 2a)$ and $y = y_1$.

Hence co-ordinates of F are $(x_1 - 2a, y_1)$

Similarly solving (4) and (6), we get $x = (x_1 + 2a)$, $y = y_1$.

Hence co-ordinates of E are $(x_1 + 2a, y_1)$

And solving (5) and (6), we get $(x = -x_1, y = -y_1)$.

Therefore the co-ordinates f D are $(-x_1, -y_1)$.

The equation of the lines passing through A and D will be

$$y + y_1 = \frac{y_1 + y_1}{x_1 + x_1}\,(x + x_1) \text{ or } yx_1 = xy_1. \quad ...(7)$$

The equation of the line passing through B and E will be

$$y - 0 = \frac{y_1 - 0}{x_1 + 2a + a}\,(x + a)$$

or $xy_1 - (3 + x_1)\,y + ay_1 = 0$. ...(8)

The equation of the line passing through C and F will be

$$y - 0 = \frac{y_1 - 0}{x_1 - 2a - a}\,(x - a)$$

or $xy_1 + (3a - x_1)\, y - ay_1 = 0.$ ...(9)

Solving (8) and (9), we get $x = \frac{x_1}{3}, y = \frac{y_1}{3}$.

Therefore the co-ordinates of the point G where the lines BE and CF meet, are $\left(\frac{x_1}{3}, \frac{y_1}{3}\right)$.

We find that the equation is satisfied by the co-ordinates of G i.e. $\left(\frac{x_1}{3}, \frac{y_1}{3}\right)$; hence the line AD passes through G also.

Hence AD, BE and CF meet in a point. **Proved.**

Example 159:

Find the bisectors of the angles between the straight lines

$$y - b = \frac{2m}{1 - m^2}(x - a) \text{ and } y - b = \frac{2m'}{1 - m'^2}(x - a).$$

Solution:

The equation of the given straight lines are

$$y - b = \frac{2m}{1 - m^2}(x - a)$$

or $$2mx - y(1 - m^2) + (b(1 - m^2) - 2am) = 0. \quad ...(1)$$

and $$y - b = \frac{2m'}{1 - m'^2}(x - a)$$

or $$2m'x - y(1 - m'^2) + (b(1 - m'^2) - 2am') = 0. \quad ...(2)$$

The equation of the bisectors are

$$\frac{2mx - y\left(1 - m^2\right) + \left\{b\left(1 - m^2\right) - 2am\right\}}{\sqrt{\left\{(2m)^2 + \left(1 - m^2\right)\right\}}}$$

$$= \pm \frac{2m'x - y\left(1 - m^2\right) + \left\{b\left(1 - m^2\right) - 2am'\right\}}{\sqrt{\left\{(2m')^2 + \left(1 - m'^2\right)\right\}}}$$

or
$$\frac{2mx - y\left(1 - m^2\right) + \left\{b\left(1 - m^2\right) - 2am\right\}}{1 - m^2}$$

$$= \pm \frac{2m'x - y\left(1 - m'^2\right) + \left\{b\left(1 - m'^2\right) - 2am'\right\}}{1 - m'^2}$$

Taking +ve sign, we get

$$2x\,(m + mm'^2 - m' - m'm^2) - y\,\{(1 - m^2)(1 + m'^2)$$
$$- (1 - m'^2)(1 + m^2) + b\,\{(1 - m^2)(1 + m'^2) - (1 - m'^2)$$
$$(1 + m^2) - 2a\,\{m + mm'^2 - m' - m'm^2\} = 0$$

or $2x\,(m - m')(1 - mm') - y.2\,(m' - m)(m' + m)$
$$+ 2b\,(m' - m)(m' + m) - 2a\,(m - m')(1 - mm') = 0$$

or $(1 - mm')(x - a) + (m + m')(y - b) = 0$

{taking 2 (m – m') common}

or $(y - b)(m + m') + (x - a)(1 + mm') = 0$...(4)

Similarly taking –ve sign in equation No. (3), we get

$$\frac{2mx - y\left(1 - m^2\right) + \left\{b\left(1 - m^2\right) - 2am\right\}}{1 - m^2}$$

$$= -\left[\frac{2m'x - y\left(1 - m'^2\right) + \left\{b\left(1 - m'^2\right) - 2am'\right\}}{1 - m'^2}\right]$$

Similarly as before, we get

$(y - b)(1 - mm') - (x - a)(m + m') = 0.$ ...(5)

Therefore equations No. (4) and (5) are the required equations. **Ans.**

Example 160:

One side of a square is inclined to the axis of x at an angle α and one of its extremities is at the origin; prove that the equations to its diagonals are $y(\cos\alpha - \sin\alpha) = x(\sin\alpha + \cos\alpha)$ and $y(\sin\alpha + \cos\alpha) + x(\cos a - \sin a) = a$, where a is the length of the side of the square

Solution:

Let OA be the one side of the square, inclined at an angle of α to CX and of length a. If B and C are the other vertices of the squares, OB is one of the diagonals.

Now $\angle BOX = \angle BCA$
$+ \angle AOX = (45° + \alpha)$.

$\therefore$ Slope of OB $= \tan(45°+\alpha)$

$$= \frac{\tan 45° + \tan \alpha}{1 - \tan 45° \tan \alpha}$$

$$= \frac{1 + \tan \alpha}{1 - \tan \alpha} = \frac{\cos \alpha + \sin \alpha}{\cos \alpha - \sin \alpha}.$$

Y
B
D
90°
A(a cos α
a sin α)
45°
α
0
N
X

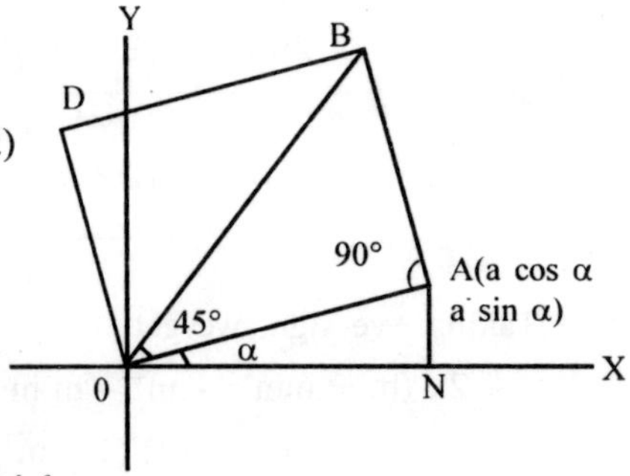

Equation of the line OB, which passes through the origin (0, 0) and has a slope given by (1), will be

$$y - 0 = \frac{\cos \alpha + \sin \alpha}{\cos \alpha - \sin \alpha}(x - 0)$$

or $\quad y(\cos \alpha - \sin a) = x(\cos \alpha + \sin \alpha)$. **Ans.**

The other diagonal AC will be perpendicular to OB.

Hence the slope of AC will be $= -\dfrac{\cos \alpha - \sin \alpha}{\cos \alpha + \sin \alpha}$, ...(2)

$(\because m_1 \times m_1 = -1)$.

Again if AN be the perpendicular on axis of x from A, ON $= a \cos \alpha$ and AN $= a \sin \alpha$; hence co-ordinates of A are $(a \cos \alpha, a \sin \alpha)$.

Equation of the line passing through A $= (a \cos \alpha, a \sin \alpha)$ and having slope given by (2) will be

$$y - a \sin a = -\frac{\cos \alpha - \sin \alpha}{\cos \alpha + \sin \alpha}(x - a \cos \alpha)$$

or $\quad y(\cos \alpha + \sin \alpha) + x(\cos \alpha - \sin \alpha)$

$$= a \sin \alpha (\cos \alpha + \sin \alpha) + a \cos \alpha (\cos \alpha - \sin \alpha)$$

$$= a \sin^2 \alpha + a \cos^2 \alpha,$$

$$= a(\sin^2 \alpha + \cos^2 \alpha) = a.$$

$\therefore$ Required equation is $y(\cos \alpha + \sin \alpha) + x(\cos \alpha - \sin \alpha) = a$. **Ans.**

Example 161:

Find the equations to the straight lines bisecting the angles between the following pairs of straight lines, placing first the bisector of the angle in which the origin lies $2x + y = 4$ and $y + 3x = 5$.

Solution:

The given equations are

$$2x + y = 4 \quad \text{or} \quad -2x - y + 4 - 0 \qquad ...(1)$$

and $$y + 3x = 5 \quad \text{or} \quad -3x - y + 5 = 0, \qquad ...(2)$$

The equation of the bisector of that angle in which the origin lies will be

$$\frac{-2x - y + 4}{\sqrt{\{(-2)^2 + (-1)^2\}}} = + \frac{-3x - y + 5}{\sqrt{\{(-3)^2 + (-1)^2\}}}$$

or $$\frac{-2x - y + 4}{\sqrt{5}} = + \frac{-3x - y + 5}{\sqrt{10}}$$

or $$x\,(2\sqrt{2} - 3) + y\,(\sqrt{2} - 1) = 4\sqrt{2} - 5. \qquad ...(3)$$

The other equation will be $\dfrac{-2x - y + 4}{\sqrt{5}} = - \dfrac{-3x - y + 5}{\sqrt{10}}$

or $$x\,(2\sqrt{2} - 3) + y(\sqrt{2} - 1) = 4\sqrt{2} - 5. \qquad ...(4)$$

Hence the required equations are (3) and (4). **Ans.**

Example 162:

Find the equations to the bisectors of the internal angles of the triangle the equations of whose sides are respectively $3x + 5y = 15$, $x + y = 4$, and $2x + y = 6$.

Solution:

The equations of the sides of the triangle are given as

$$3x + 5y - 15 = 0 \qquad ...(1)$$

$$x + y - 4 = 0 \qquad ...(2)$$

and $$2x + y - 6 = 0 \qquad ...(3)$$

Let the lines represent AB, BC and CA respectively.

Solving (1) and (3), the co-ordinates of A are $\left(\frac{15}{7}, \frac{12}{7}\right)$.

Solving (1) and (3), the co-ordinates of B are $\left(\frac{5}{2}, \frac{3}{2}\right)$.

Solving (2) and (3), the co-ordinates of C are (2, 2).

Substituting the co-ordinates of C = (2, 2) in equation (1) of AB, we get $3 \times 2 + 5 \times 2 - 15$ = +ve; hence multiplying (1) by 1, we get

$$3x + 5y - 15 = 0. \qquad ...(4)$$

Substituting the co-ordinates of A in equation (2) of BC we get $\frac{15}{7}, \frac{12}{7}$ $- 4 = -$ve. So multiplying (2) by -1, we get

$$-x - y + 4 = 0. \qquad ...(5)$$

Substituting the co-ordinates of B in equation (3) of AC, we get $2 \times \frac{5}{2}, \frac{3}{2} - 6 = -$ve. Hence multiplying (3) by 1, we get

$$2x + y - 6 = 0. \qquad ...(6)$$

Therefore the bisector of $\angle ABC$ will be the bisector of the lines given by (4) and (6); so

$$\frac{3x + 5y - 15}{\sqrt{(9 + 25)}} = + \frac{-x - y + 4}{\sqrt{(1 + 1)}}$$

or $$\frac{3x + 5y - 15}{\sqrt{(34)}} = + \frac{-x - y + 4}{\sqrt{(2)}}$$

or $$(3x + 5y - 15) = (-x - y + 4) \times \sqrt{17}$$

or $$x [3 + \sqrt{17}] + y [5 + \sqrt{17}] = 15 + 4 \sqrt{17}.$$ **Ans.**

Similarly the equation of the bisector of $\angle ACB$ will bee the bisector of the angle between the lines given by (5) and (6)

or $$\frac{-x - y + 4}{\sqrt{(2)}} = \frac{2x + y - 6}{\sqrt{(4 + 1)}} = \frac{2x + y - 6}{\sqrt{5}}$$

or $$\sqrt{5} (-x - y + 4) = \sqrt{2} (2x + y - 6)$$

or $$\sqrt{(10)} (-x - y + 4) = 2 (2x + y - 6)$$

or $$x [x + \sqrt{(10)}] + y [2 + \sqrt{(10)}] = 4 \sqrt{10} + 12.$$ **Ans.**

And the equation of the bisector of $\angle BAC$ is the bisector of the angle between the lines (4) and (6); so

$$\frac{3x + 5y - 15}{\sqrt{(34)}} = \frac{2x + y - 6}{\sqrt{5}}$$

or $$x [2\sqrt{(34)} - 3\sqrt{5}] + y [\sqrt{(34)} - 5\sqrt{5}] = 6\sqrt{(34)} - 15\sqrt{5}.$$ **Ans.**

Example 163:

Find the equations to the bisectors of the internal angles of the triangle the equations of whose sides are respectively

$$3x + 4y = 6,$$

$$12x - 5y = 3,$$

and $$4x - 3y + 12 = 0.$$

Solution:

The equations of 3 sides of the triangle are given as

$$3x + 4y = 6 \qquad ...(1)$$

$$12x - 5y = 3 \qquad ...(2)$$

and $$4x - 3y + 12 = 0 \qquad ...(3)$$

Let the three lines represent AB, BC and CA respectively.

Solving (1) and (3), we get the co-ordinates of A as $\left(-\frac{6}{5}, \frac{12}{5}\right)$.

Solving (1) and (2), we get the co-ordinates of B as $\left(\frac{2}{3}, 1\right)$.

Solving (2) and (3), we get the co-ordinates of C as $\left(\frac{33}{8}, \frac{39}{4}\right)$.

Substituting the co-ordinates of C = $\left(\frac{33}{8}, \frac{39}{4}\right)$ in the equation of the opposite side AB, we get $3 \times \frac{33}{8} + 4 \times \frac{39}{4} - 6$ = +ve quantity.

Multiplying the equation by 1, we get $3x + 4y - 6 = 0$. ...(4)

Substituting the co-ordinates of the point A $\left(-\frac{6}{5}, \frac{12}{5}\right)$ in the equation of BC, we get $12 \times -\frac{6}{5} - 5 \times -\frac{12}{5} - 3$ = –ve quantity.

Hence multiplying the equation (2), i.e. equation of BC by – 1, we get

$$-12x + 5y + 3 = 0. \qquad ...(5)$$

Substituting the co-ordinates of B in the equation (3) of AC, we get $4 \times \frac{2}{3} - 3 \times 1 + 12$ = +ve quantity. Hence multiplying the equation (3) by 1, we get

$$4x - 3y + 12 = 0. \qquad ...(6)$$

Now the equation of the bisector of ∠ ABC will be the bisector of the lines given by (4) and (5)

$$\therefore \qquad \frac{3x + 4y - 6}{\sqrt{(9 + 16)}} = + \frac{-12x + 5y + 3}{\sqrt{(144 + 25)}}$$

or $$\frac{3x + 4y - 6}{5} = + \frac{-12x + 5y + 3}{13}$$

or $$99x + 27y - 93 = 0$$

or $$33x + 9y - 31 = 0.$$ **Ans.**

Similarly the equation of the bisector of $\angle ACB$ will be the bisector of the angle between (5) and (6).

$$\frac{-12x + 5y + 3}{\sqrt{(144 + 25)}} = + \frac{4x - 3y + 12}{\sqrt{(16 + 9)}}$$

or $$\frac{-12x + 5y + 3}{13} = \frac{4x - 3y + 12}{15}$$

or $$112x - 64y + 141 = 0.$$ **Ans.**

and the bisector of $\angle BAC$ will be the bisector of (4) and (6).

$$\frac{3x + 4y - 6}{5} = + \frac{4x - 3y + 12}{15}$$

or $$x - 7y + 18 = 0.$$ **Ans.**

Example 164:

Find the direction in which a straight line must be drawn through the point (1,2), so that its point of intersection with the line $x + y = 4$ may be at a distance $\sqrt{6}/3$ from this point.

Solution:

Let the equation of the required line which passes through $P = (1, 2)$ be inclined at an angle of θ to x-axis.

Hence its equation may be given by $\frac{x - 1}{\cos\theta} = \frac{y - 2}{\sin\theta} = r.$ ...(1)

where r is the distance of any point on the line from (1, 2).

Let this line meet the given line $x + y = 4$ ...(2)

in any point Q. The $PQ = \sqrt{6}/3$ (given).

Let us suppose $r = \sqrt{6}/3$ and substituting in (1), we get

$$\frac{x - 1}{\cos\theta} = \frac{y - 2}{\sin\theta} = \frac{\sqrt{6}}{3}$$

or $$x = \left(\frac{\sqrt{6}}{3}\cos\theta + 1\right);\ y = \left(\frac{\sqrt{6}}{3}\sin\theta + 2\right)$$

Hence the co-ordinates of Q will be

$$\left[\frac{\sqrt{6}}{3}\cos\theta + 1,\ \frac{\sqrt{6}}{3}\sin\theta + 2\right].$$

As Q lies on the given line (2) also, the co-ordinates will satisfy the equation (2); hence

$$\frac{\sqrt{6}}{3}\cos\theta + 1 + \frac{\sqrt{6}}{3}\sin\theta + 2 = 4$$

or $$\sqrt{6}\,(\cos\theta + \sin\theta) = 3$$

or $$\cos\theta + \sin\theta = \frac{3}{\sqrt{6}} = \frac{\sqrt{3}}{\sqrt{2}}$$

or $$\frac{1}{\sqrt{2}}\cos\theta + \frac{1}{\sqrt{2}}\sin\theta = \frac{\sqrt{3}}{2}$$ (multiplying throughout by $\frac{1}{\sqrt{2}}$)

or $$\cos\frac{\pi}{4}\cos\theta + \sin\frac{\pi}{4}\sin\theta = \frac{\sqrt{3}}{2}$$

or $$\cos(\theta - 45°) = \frac{\sqrt{3}}{2}.$$

Hence either $\theta - 45° = 30°$; $\therefore \theta = 75°$

or $$\theta - 45° = -30°$$

or $$\theta = 15°.$$

Therefore the required angles are 75° or 15°. **Ans.**

Example 165:

Find the equation to the straight line passing through the foot of the perpendicular from the point (h, k) upon the straight line Ax + By + C = 0, and bisecting the angles between the perpendicular and the given straight line.

Solution:

Equation of the given line is

$$Ax + By + C = 0. \qquad \ldots(1)$$

Equation of any line perpendicular to (1) will be $Bx - Ay = \lambda$, where λ is any constant. As the perpendicular line passes through (h, k), hence it will satisfy (2). So Bh – Ak = l. Substituting in (2), we get the equation of the line perpendicular to (1) and passing through (h, k) as

$$Bx - Ay = Bh - Ak \qquad ...(2)$$

or $$Bx - Ay - Bh + Ak = 0. \qquad ...(3)$$

Equations of the bisectors of the angles between the lines given by (1) and (3) will be

$$\frac{Ax + By + C}{\sqrt{(A^2 + B^2)}} = \pm \left[\frac{Bx - Ay - Bh + Ak}{\sqrt{(B^2 + A^2)}} \right]$$

or $$Ax + By + C = \pm [B(x - h) - A(y - k)]$$

or $$A(y - k) - B(x - h) = \pm (Ax + By + C).$$ **Ans.**

②

Equations Representing Two or More Straight Lines

TWO OR MORE STRAIGHT LINES

Angle Between Two Straight Lines

To find the angle between given straight lines

$$y = m_1x + c_1$$

and $$y = m_2x + c_2.$$

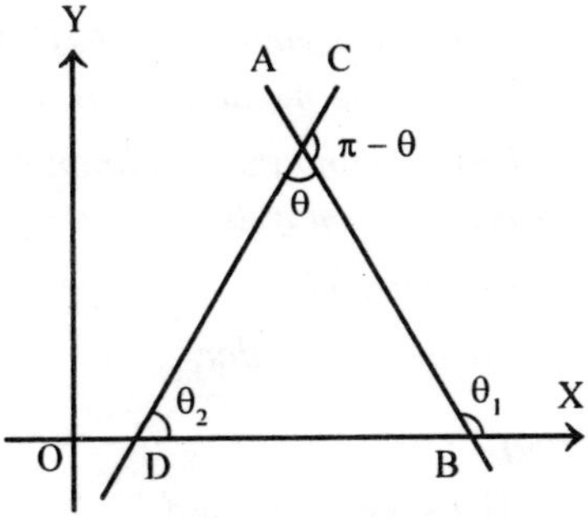

Proof :

Let AB and CD be two given straight lines whose equations are given by

$$y = m_1x + c_1 \text{ and } y = m_2x + c_2$$

respectively. Let P be the point of intersection of AB and CD, so that if θ_1 and θ_2 are the angles they make with the positive direction of x-axis then

$$\tan\theta_1 = m_1 \text{ and } \tan\theta_2 = m_2 \qquad ...(i)$$

Let θ be the required angle between then we have two lines then

$$\theta_1 = \theta + \theta_2$$

or $$\theta = \theta_1 - \theta_2.$$

$$\therefore \quad \tan\theta = \tan(\theta_1 - \theta_2)$$

$$= \frac{\tan\theta_1 - \tan\theta_2}{1+\tan\theta_1\tan\theta_2}$$

$$= \frac{m_1 - m_2}{1+m_1m_2} \qquad ...(ii)$$

$$\therefore\ \theta = \tan^{-1} = \frac{m_1 - m_2}{1 + m_1 m_2}$$

Notes. 1. *If ϕ is the other angle between the two given lines then*

$$\phi = \pi - 0$$

or $\quad \tan \phi = \tan (\pi - \theta) = -\tan \theta$

$$= \frac{m_1 - m_2}{1 + m_1 m_2}. \qquad \text{...(iii)}$$

$\therefore$ Combining the results (ii) and (iii) we have

$$\tan\theta = \pm \frac{m_1 - m_2}{1 + m_1 m_2}$$

which is the complete *angle formula* between two give lines.

2. *If* $\frac{m_1 - m_2}{1 + m_1 m_2}$ *is +ve, then* $\tan^{-1}\frac{m_1 - m_2}{1 + m_1 m_2}$ *is the acute angle between the lines and if* $\frac{m_1 - m_2}{1 + m_1 m_2}$ *is negative, then* $\tan^{-1}\frac{m_1 - m_2}{1 + m_1 m_2}$ *is the obtuse angle between the given lines.*

3. *To numerical examples, if the angle between two lines is required, we discard the negative sign and find the acute angle between the lines.*

Cor. *To find the condition that the two straight lines $y = m_1 x + c_1$ and $y = m_2 x + c_2$ may be*

(i) parallel,

(ii) perpendicular.

Proof:

The given lines are

$$y = m_1 x + c_1$$

and $\quad y = m_2 x + c_2.$

If θ is the angle between the lines, then

$$\tan\theta = \frac{m_1 - m_2}{1 + m_1 m_2}.$$

(i) If the lines are parallel, then $\theta = 0^\circ$

$\therefore \quad \tan \theta = \tan 0^\circ = 0$

$$\therefore \qquad 0 = \frac{m_1 - m_2}{1 + m_1 m_2}$$

$\therefore \quad m_1 - m_2 = 0$

or $\quad m_1 = m_2$

which is the required condition of parallelism of two lines.

Hence, two given lines will be parallel if their slopes are equal and conversely.

(ii) If the lines are perpendicular to each other, then $\theta = 90^o$.

$\therefore \tan\theta = \tan 90^o = \infty$.

$$\therefore \quad \frac{m_1 - m_2}{1 + m_1 m_2} = \infty$$

$$\Rightarrow \quad \frac{m_1 - m_2}{1 + m_1 m_2} = \frac{1}{\infty} = 0$$

$$\Rightarrow \quad 1 + m_1 m_2 = 0$$

$$\Rightarrow \quad m_1 m_2 = -1$$

which is the required condition for two lines to be perpendicular.

Hence, two given lines are perpendicular if the product of their slopes is equal to –1 and conversely. i.e.

$$m_1 \times m_2 = -1$$

To find the angle between the straight lines

$$a_1x + b_1y + c_1 = 0$$

and $\quad a_2x + b_2y + c_2 = 0.$

Proof:

Let m_1 and m_2 be the slopes of the two given lines, then we have

$$a_1x + b_1y + c_1 = 0$$

and $\quad a_2x + b_2y + c_2 = 0.$

Then we have $m_1 = \frac{a_1}{b_1}$ and $m_2 = \frac{a_2}{b_2}$.

If θ is the angle between the lines, then

$$\tan\theta = \frac{m_1 - m_2}{1 + m_1 m_2} = \frac{\left(\frac{a_2}{b_2}\right) - \left(\frac{a_1}{b_1}\right)}{1 + \left(-\frac{a_1}{b_1}\right)\left(-\frac{a_2}{b_2}\right)}$$

$$\Rightarrow \frac{\frac{a_2}{b_2} - \frac{a_1}{b_1}}{1 + \frac{a_1 a_2}{b_1 b_2}} = \frac{a_2 b_1 - b_2 a_1}{a_1 a_2 + b_1 b_2}$$

$$\therefore \theta = \tan^{-1}\left(\frac{a_2 b_1 - b_2 a_1}{a_1 a_2 + b_1 b_2}\right)$$

Cor. : *Conditions for the lines*

$$a_1 x + b_1 y + c_1 = 0$$

and $$a_2 x + b_2 y + c_2 = 0$$

may be (i) parallel, (ii) perpendicular. are given as

(i) If the lines are parallel, then $\theta = 0^\circ$

$\therefore \tan\theta = \tan 0^\circ = 0$

$$\therefore \quad 0 = \frac{a_2 b_1 - b_2 a_1}{a_1 a_2 + b_1 b_2}$$

$$\Rightarrow \quad a_2 b_1 - a_1 b_2 = 0$$

$$\Rightarrow \quad \frac{a_1}{a_2} = \frac{b_1}{b_2}$$

(ii) If the lines are perpendicular, then $\theta = 90^\circ$

$\therefore \tan\theta = \tan 0^\circ = \infty$

$$\therefore \quad \frac{a_2 b_1 - a_1 b_2}{a_2 b_1 + a_1 b_2} = \infty$$

$$\Rightarrow \quad \frac{a_1 a_2 + b_1 b_2}{a_2 b_1 - a_1 b_2} = \frac{1}{\infty} = 0$$

$$\Rightarrow \quad a_1 a_2 + b_1 b_2 = 0.$$

SOME SYSTEM OF STRAIGHT LINES

In this section, we shall give the general formula for writing the equation of straight line (i) parallel to a given straight line, (ii) perpendicular to given straight line.

(i) Lines perpendicular to $ax + by + c = 0$

The slope of this line is $\left(\frac{-a}{b}\right)$, therefore the slope of the line which are perpendicular to the line will be $\left(\frac{-1}{-a/b}\right) = \frac{b}{a}$. Hence the equation of line having b/a as slope can be written as

$$y = \left(\frac{b}{a}\right) x + \lambda$$

$$\Rightarrow \qquad bx - ay + \lambda = 0.$$

where λ is any arbitrary constant.

Hence the equation of a line perpendicular to a given line is obtained by (i) interchanging the co-efficients of x and y, (ii) changing the sign of one of them, and (iii) adding the constant term whose value can be obtained under the given condition.

(ii) Lines parallel to $ax + by + c = 0$.

Since the slopes of parallel lines are equal, the slope of the new line parallel to given line will be $(-a/b)$. Equation of the line with this slope is

$$y = \left(\frac{-a}{b}\right) x + \lambda$$

$$\Rightarrow \qquad ax + by + \lambda = 0,$$

where λ is any arbitrary constant.

Hence, the equation of a straight line parallel to a given straight line is obtained by changing the constant term whose value can be obtained subject to given conditions.

Example 1:

Find the area of the triangle formed. By the lines whose equations are $y = m_1x + c_1$, $y = m_2x + c_2$ and $x = 0$.

Solution:

The equation of the sides of a triangle are

$$y = m_1x + c_1 \qquad \text{...(i)}$$

$$y = m_2x + c_2 \qquad \text{...(ii)}$$

$$x = 0. \qquad \text{(iii)}$$

The point of intersection of (i) and (ii) is obtained by solving these two equations simultaneously. This gives

$$\left(\frac{c_2 - c_1}{m_1 - m_2}, \frac{c_2m_1 - c_1m_2}{m_1 - m_2}\right).$$

Similarly, $(0, c_1)$ is the point of intersection (i) and (iii) and $(0, c_2)$ is the point of intersection of (ii) and (iii). Hence the required area of the triangle is

$$\frac{1}{2}\begin{vmatrix} 0 & c_1 & 1 \\ 0 & c_2 & 1 \\ \frac{c_2-c_1}{m_1-m_2} & \frac{c_2m_1-c_1m_2}{m_1-m_2} & 1 \end{vmatrix}$$

$$=\frac{1}{2}\left(\frac{c_2-c_2}{m_1-m_2}\right)(c_1-c_2)$$

$$=\frac{1}{2}\frac{(c_2-c_2)^2}{m_1-m_2}$$

Example 2:

Show that the equation of the straight line passing through the point $(a \cos^3 \theta, a \sin^3 \theta)$ and perpendicular to the line

$\alpha \sec \theta + y \operatorname{cosec} \theta = a$ is

$x \cos \theta - y \sin \theta = \cos 2\theta$.

Solution:

The equation of the straight line perpendicular to the given straight line is given as

$x \operatorname{cosec} \theta - y \sec \theta + \lambda = 0$ ...(i)

It passes through the point $(a \cos^3 \theta, a \sin^3 \theta)$. Then we have

$a \cos^3 \theta - \operatorname{cosec} \theta - a \sin^3 \theta \sec \theta + \lambda = 0$

$$\Rightarrow \lambda = a\left(\frac{\sin^3\theta}{\cos\theta}-\frac{\cos^3\theta}{\sin\theta}\right)=\frac{-a\cos 2\theta}{\cos\theta\sin\theta}$$

Substituting λ in (i) we have

$$x \operatorname{cosec} \theta - y \sec \theta = \frac{a\cos 2\theta}{\cos\theta\sin\theta}$$

$x \cos \theta - y \sin \theta \, a \cos 2\theta = 0$

Example 3:

For what value of x, the line joining (–5, 7) and (0, – 2) is perpendicular to the line joining (1, – 3) and (4, x).

Solution:

Equation of the line joining (–5, 7) and (0, –2) is given as

$$y - 7 = \frac{7+2}{-5-0}(x+5)$$

$$\Rightarrow y - 7 = \left(\frac{9}{-5}\right)(x+5). \qquad \text{...(i)}$$

Equation of the line joining (1, –3) and (4, x) is

$$y + 3 = \frac{-3-x}{1-4}(x-1). \qquad \text{...(ii)}$$

The lines will be perpendicular if the product of the slopes is (–1). This gives

$$\left(\frac{9}{-5}\right)\left(\frac{3+x}{3}\right) = -1$$

$$\Rightarrow \qquad x = -\,4/3.$$ **Ans.**

LENGTH OF PERPENDICULAR

To find the length of the perpendicular let fall from a given point upon a given straight line.

Case I. *When the equation of the given straight line is*

$x \cos \alpha + y \sin \alpha - p = 0$

Let AB be a straight line and $P(x_1, y_1)$ be any point. From P draw a line CD parallel to the given line and let P' (OF) be the perpendicular distance from O on CD. The equation of CD is given by

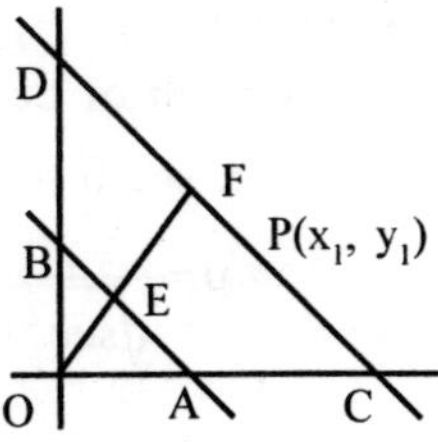

$x \cos \alpha + y \sin \alpha = p'$

This line passes through the point $P(x_1, y_1)$.

$\therefore x_1 \cos \alpha + y_1 \sin \alpha = p'$

FE = perpendicular distance from P on the line AB

= FO – EO

= p' – p

$= x_1 \cos \alpha + y_1 \sin \alpha = p.$

Hence, the length of the perpendicular from a point (x_1, y_1) on the line $x \cos \alpha + y \sin \alpha = p$ is given as

$x_1 \cos \alpha + y_1 \sin \alpha - p$

Case II. *When the equation of the given straight line is given by* $ax + by + c = 0$.

The equation of the given straight line is

$ax + by + c = 0$

Reducing it to perpendicular form by dividing throughout by $\sqrt{a^2 + b^2}$ we have

$$\frac{ax}{\sqrt{a^2+b^2}} + \frac{by}{\sqrt{a^2+b^2}} + \frac{c}{\sqrt{a^2+b^2}} = 0$$

Now applying the result obtained in case I, we find the length of the perpendicular from the point (x_1, y_1) to the given line $ax + by + c$ is

$$\frac{ax_1 + by_1 + c}{\sqrt{a^2+b^2}}$$

Example:

If p and p' are the perpendicular from the origin on the straight lines whose equations are

$$x \sec \theta - y \text{ cosec } \theta = a$$

and $x \cos \theta + y \sin \theta = a \cos q2\theta$,

prove that $4p^2 + p'^2 = a^2$.

Solution:

Length of perpendicular from (0, 0) to the line

$x \sec \theta - y \text{ cosec } \theta - a = 0$ is given as

$$p = \frac{a}{\sqrt{\sec^2 \theta + \text{cosec}^2 \theta}}$$

$$\Rightarrow p^2 = \frac{a^2}{\frac{1}{\cos^2 \theta} + \frac{1}{\sin^2 \theta}} = \frac{a^2 \sin^2 \theta \cos^2 \theta}{\sin^2 \theta + \cos^2 \theta}$$

$\therefore p^2 = a^2 \sin^2 \theta \cos^2 \theta \qquad (\because \sin^2 \theta + \cos^2 \theta = 1)$

Also length of perpendicular from (0, 0) to the line is

$x \cos \theta + y \sin \theta - a \cos 2\theta = 0$ is

$$p' = \frac{-a\cos 2\theta}{\sqrt{\cos^2\theta + \sin^2\theta}} = -a\cos 2\theta$$

$\therefore p'^2 = a^2 \cos^2 2\theta$

$\therefore 4p^2 + p'^2 = 4a^2 \sin^2\theta \cos^2\theta + a^2 \cos^2.\ 2\theta$

$= a^2(\sin^2 2\theta + \cos^2 2\theta) = a^2$

POSITION OF POINT WITH RESPECT TO A LINE

To show that two points (x_1, y_1) and (x_2, y_2) are on the same side of the line $ax + by + c = 0$ if the expression $ax_1 + by_1 + c$ and $ax_2 + by_2 + c$ have the same sign and on the opposite sides if the expressions are of opposite sign.

Proof:

Let AB be the straight line $ax + by + c = 0$ and $P(x_1, y_1)$ and $Q(x_2, y_2)$ be two points. Let PS and QS' be the perpendiculars on x-axis meeting the line at R and R'. Since R and R' have the same abscissas as P and Q respectively the ço-ordinates of and R' can be taken as (x_1, y_1) and $R(x_2, y_2')$ respectively

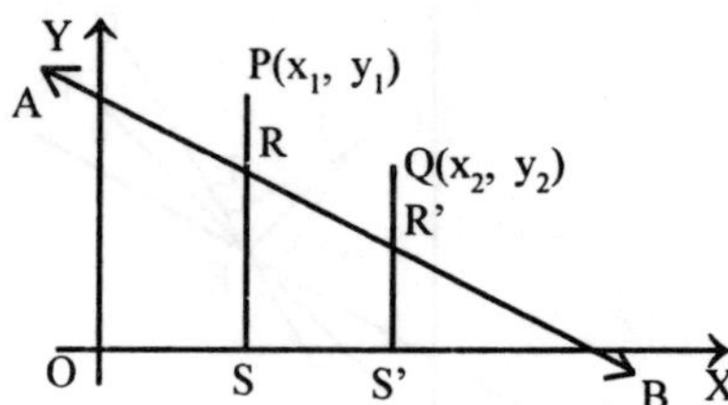

Case I. P, Q lie on same side

(i) above the line

$\Rightarrow \quad y_1 > y_1'$ and $y_2 > y_2'$

Also $ax_1 + by_1' + c = 0$ and $ax_2 + by_2' + c = 0$ (Since R and R' lie on AB)

$\therefore \quad ax_1 + by_1 + c > ax_1 + by_1' + c = 0$

and $\quad ax_2 + by_2 + c > ax_2 + by_2' + c = 0$

$\therefore$ Both the expressions have the same sign.

(ii) below the line.

In this case $y_1 < y_1'$ and $y_2 < y_2'$ both the expressions $ax_1 + by_1 + c$ and $ax_2 + by_2 + c$ will have negative sign. Thus in both the cases the expressions have the same sign.

Case II. P, Q lie on opposite side

$\Rightarrow \quad y_1 < y_1', \; y_2 < y_2'$

or $\quad y_1 < y_1', \; y_2 < y_2'$

$\Rightarrow \quad ax_1 + by_1 + c > ax_1 + by_1' + c > 0$

and $\quad ax_2 + by_2 + c < ax_2 + by_2' + c < 0$

or $\quad ax_1 + by_1 + c < ax_1 + by_1' + c < 0$

and $\quad ax_2 + by_2 + c > ax_2 + by_2' + c > 0$

Thus in both the possibilities the expressions have opposite signs.

Corollary 1. The point (x_1, y_1) and $(0, 0)$ lie on the same side of the line $ax + by + c = 0$ if $ax_1 + by_1 + c > 0$.

BISECTORS OF ANGLES

To find the equation of the straight lines which bisect the angles between two given straight lines.

Let AB and CD be given lines intersecting at E. Let their equations be given by

$$a_1x + b_1y + c_1 = 0 \quad \text{...(i)}$$

$$a_2x + b_2y + c_2 = 0 \quad \text{...(ii)}$$

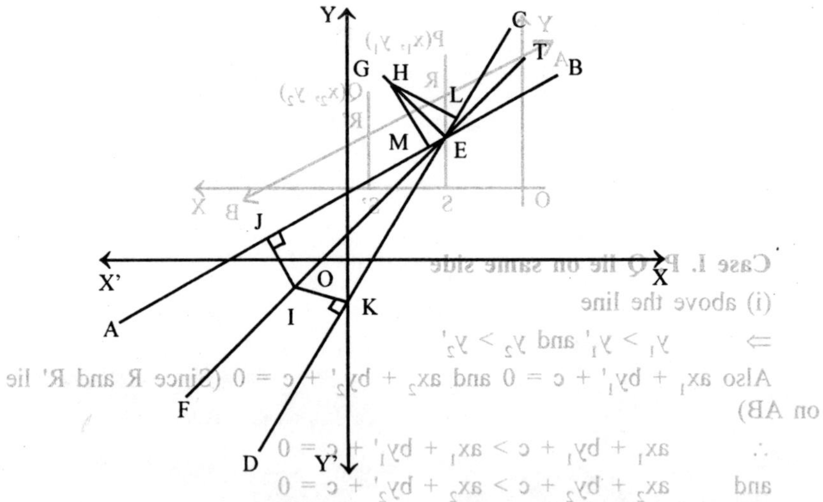

Let EF and EG be the bisectors of angles $\angle AED$ and $\angle CEA$. Take any two points I and H on EF and EG respectively. From I draw perpendiculars IJ and IK on AB and CD respectively. From H draw HL and HM perpendiculars on CD and AB respectively.

Points I and H are on the same side of CD

⇒ IK and HL have the same sign.

Points I and H are on the opposite sides of AB

⇒ IJ and HM have opposite signs.

⇒ If IJ and IK have same sign then HL and HM have opposite signs and vice versa.

By simple properties of a triangle

IJ = IK

HL = HM.

If (x, y) denote the co-ordinate of I or H then using the perpendicular distance formula we have the equation of EF as

$$\frac{a_1x+b_1y+c_1}{\sqrt{a_1^2+b_1^2}} = \frac{a_2x+b_2y+c_2}{\sqrt{a_2^2+b_2^2}} \quad \text{...(i)}$$

and the equation of EG

$$\frac{a_1x+b_1y+c_1}{\sqrt{a_1^2+b_1^2}} = -\frac{a_2x+b_2y+c_2}{\sqrt{a_2^2+b_2^2}} \quad \text{...(ii)}$$

This gives the required equation of the bisectors as

$$\frac{a_1x+b_1y+c_1}{\sqrt{a_1^2+b_1^2}} = \pm\frac{a_2x+b_2y+c_2}{\sqrt{a_2^2+b_2^2}}$$

DISTINCTION BETWEEN THE TWO BISECTORS WITH THE HELP OF THE ORIGIN

Let the equations of the two lines be given by

$$u \equiv a_1x + b_1y + c_1 = 0$$

and $$v \equiv a_2x + b_2y + c_2 = 0$$

where c_1 and c_2 are both positive.

Then the origin lies on the positive side of each of the lines.

Let $P_1(x_1, y_1)$ be any point on the bisector which contains the origin. Since P_1 and O lie on the same side of each line the perpendicular from P_1 on u = 0, v = 0 must be positive. Also since c_1 and c_2 are positive, there fore $a_1x_1 + b_1y_1 + c_1$ and $a_2x_2 + b_2y_2 + c_2$ are also positive. Hence the bisector which contains the origin is

$$\frac{a_1x+b_1y+c_1}{\sqrt{a_1^2+b_1^2}} = \frac{a_2x+b_2y+c_2}{\sqrt{a_2^2+b_2^2}}$$

LINES THROUGH THE POINT OF INTERSECTION OF TWO STRAIGHT LINES

Show that the equation of any straight line passing through the intersection of straight line $u = 0$ and $v = 0$ is $u + \lambda v = 0$, where λ is any arbitrary constant.

Proof:

Let the equation of the straight lines be given by

$$u \equiv a_1x + b_1y + c_1 = 0 \qquad \text{...(i)}$$

$$v \equiv a_2x + b_2y + c_2 = 0 \qquad \text{...(ii)}$$

Let (h, k) be the point of intersection of lines (i) and (ii). It will be shown that (h, k) lies on the line

$$a_1x + b_1y + c_1 + \lambda (a_1x + b_1y + c_1) = 0. \qquad \text{...(iii)}$$

Since (h, k) is the point of intersection of lines (i) and (ii), we have

$$a_1h + b_1k + c_1 = 0. \qquad \text{...(iv)}$$

$$a_2h + b_2k + c_2 = 0. \qquad \text{...(v)}$$

Now the expression

$$(a_1h + b_1k + c_1) + \lambda (a_2h + b_2k + c_2) = 0$$

for all values of λ.

This shows that (h, k) lies on line (iii)

Thus whatsoever be the values of λ, equation (iii) represents a straight line passing through the intersection of (i) and (ii).

CONDITION FOR THE CONCURRENCY OF THREE STRAIGHT LINES

Let the three straight lines be

$$a_1x + b_1y + c_1 = 0$$

$$a_2x + b_2y + c_2 = 0$$

$$a_3x + b_3y + c_3 = 0$$

and let these lines meet at the point (h, k). Then we have

$$a_1h + b_1k + c_1 = 0$$

$a_3h + b_2k + c_2 = 0$

$a_3h + b_3k + c_3 = 0.$

Eliminating h and k between these three equations, we get

$$\begin{vmatrix} a_1 & b_1 & c_1 \\ a_2 & b_2 & c_2 \\ a_3 & b_3 & c_3 \end{vmatrix} = 0,$$

which is the required condition for the concurrency of three straight lines.

Another test for concurrency of three lines

Let the three lines be

$A_1x + B_1y + C_1 = 0$ ...(i)

$A_2x + B_2y + C_2 = 0$ ...(ii)

$A_3x + B_3y + C_3 = 0$ ...(iii)

Definition: *If there constants p, q, r, can be found and such that*

$p(A_1x + B_1y + C_1) + q(A_2x + B_2y + C_2) + r(A_3x + B_3y + C_3) \equiv 0$ ...(iv)

Identically, (i.e., expression vanishes irrespective of the values of x and y) then the three lines must meet in a point.

Proof:

Let (α, β) be the point of intersection of (i) and (ii), then

$A_1\alpha + B_1\beta + C_1 = 0$ ...(v)

$A_2\alpha + B_2\beta + C_2 = 0$ ...(vi)

Now (iv) is an identity in x and y and hence true for all values of x and y. Putting $x = \alpha$ and $y = \beta$ in (iv), we have

$$p(A_1\alpha + B_1\beta + C_1) + q(A_2\alpha + B_2\beta + C_2) + r(A_3\alpha + B_3\beta + C_3) \equiv 0$$

$$A_3\alpha + B_3\beta + C_3 = -\frac{p}{r}(A_1\alpha + B_1\beta + C_1) - \frac{q}{r}(A_2\alpha + B_2\beta + C_2)$$

$$= -\frac{p}{r}.0 - \frac{q}{r}.\ 0 = 0 \qquad \text{[By (v) and (vi)]}$$

which show that (α, β) is also a point on the line (iii).

Hence, the three lines are concurrent.

Example 1:

Show that the lines $lx + my + n = 0$, $mx + ny + l = 0$ and $nx + ly + m = 0$ are concurrent if $l + m + n = 0$.

Solution:

The equation of the lines are

$lx + my + n = 0$

$mx + ny + l = 0$

$nx + ly + m = 0.$

The lines are concurrent if

$$\begin{vmatrix} l & m & n \\ m & n & l \\ n & l & m \end{vmatrix} = 0$$

Operating $C_1 + (C_2 + C_3)$, we have

$$\begin{vmatrix} l+m+n & m & n \\ l+m+n & n & l \\ l+m+n & l & m \end{vmatrix} = 0$$

$$\text{or } (l + m + n) \begin{vmatrix} 1 & m & n \\ 1 & n & l \\ 1 & l & m \end{vmatrix} = 0$$

which holds as $l + m + n = 0$.

Example 2:

Prove analytically that altitude of a triangle meet at a point.

Solution:

Let $A(x_1, y_1)$ $B(x_2, y_2)$ and $C(x_3, y_3)$ be the vertices of a triangle. Draw AD is perpendicular from A on BC.

Now slope of $BC = \dfrac{y_3 - y_2}{x_3 - x_2}$

$\therefore$ Slope of $AD = \dfrac{(x_3 - x_2)}{y_3 - y^2}$

$\therefore$ equation of AD is $y - y_1 = -\dfrac{x_3 - x_2}{y_3 - y_2}(x - x_1)$

which on simplification becomes

$x(x_3 - x_2) + y(y_3 - y_2) - y_1(y_3 - y_2) - x_1(x_3 - x_2) = 0$ ...(i) similarly the equation of altitudes BE and CF are

$$x(x_1 - x_3) + y(y_1 - y_3) - y_2(y_1 - y_3) - x_2(x_1 - x_3) = 0 \quad ...(ii)$$

and $$x(x_2 - x_1) + y(y_2 - y_1) - y_3(y_2 - y_1)\; x_3(x_2 - x_1) = 0 \quad ...(iii)$$

These lines will be concurrent if

$$\begin{vmatrix} x_3 - x_2 & y_3 - y_2 & y_1(Y_3 - y_2) + x_1(x_3 - x_2) \\ x_1 - x_3 & y_1 - y_3 & y_2(Y_1 - y_3) + x_2(x_1 - x_3) \\ x_2 - x_1 & y_2 - y_1 & y_3(Y_2 - y_1) + x_3(x_2 - x_3) \end{vmatrix} = 0$$

Operating $R_1 + R_2 + R_3$ we get

$$\begin{vmatrix} 0 & 0 & 0 \\ x_1 - x_3 & y_1 - y_3 & y_2(y_1 - y_3) + x_2(x_1 - x_3) \\ x_2 - x_1 & y_2 - y_1 & y_3(y_2 - y_1) + x_3(x_2 - x_3) \end{vmatrix} = 0.$$

Hence the lines are concurrent

Example 3:

A triangle is formed by the lines $y + x - 6 = 0$, $3y - x + 2 = 0$, $5x + 3y + 2 = 0$. Find the coordinates of its orthocentre.

Solution:

Let ABC be the given triangle the equations of whose sides BC, CA and AB are

$$x + y - 6 = 2 \quad ...(i)$$

$$-x + 3y + 2 = 0 \quad ...(ii)$$

$$5x - 3y + 2 = 0. \quad ...(iii)$$

A
3y – x + 2 = 0
5x – 6y + 2 = 0
M
H
B
C
L y + x – 6 = 0

The coordinates of the point of intersection of the lines AB and AC are given by

$$\frac{x}{-6-6} = \frac{y}{-10-2} = \frac{1}{-3+15}$$

$\Rightarrow x = -1, \; y = -1.$

$\therefore$ the coordinates of A are $(-1, -1)$.

The coordinates of the point of intersections of the lines BC and AB are given by

$$\frac{x}{2-18} = \frac{y}{-30-2} = \frac{1}{-3-5};$$

$\therefore$ x = 2. y = 4.

$\therefore$ the coordinates of B are (2, 4).

The slope of BC is –1 and the slope of altitude AL is 1.

$\therefore$ the equation of the line AL is

$y + 1 = 1(x + 1)$

or $x - y = 0$ ...(iv)

Slope of altitude BM is –3.

$\therefore$ the equation of the line BM is

$y - 4 = -3(x - 2)$

$\Rightarrow$ $3x + y - 10 = 0.$ ...(v)

Solving (iv) and (v), we get $x = \frac{5}{2}$ and $y = \frac{5}{2}$.

$\therefore$ the coordinates of the orthocentre H are $\left(\frac{5}{2}, \frac{5}{2}\right)$.

JOINT EQUATION FOR MORE THAN ONE STRAIGHT LINE

Consider the equation

$(a_1x + b_1y + c_1)(a_2x + b_2y + c_2) = 0.$ ...(i)

Any point (h, k) which satisfies $a_1x + b_1y + c_1 = 0$ will also satisfy equation (i) similarly, every point on the line $a_2x + b_2y + c_2 = 0$ also satisfies equation (i). Thus equation (i) is satisfied by any point on either of the given lines, and conversely, on point which does not satisfy any of the lines cannot satisfy equation (i). Thus (i) represents the joint equation of the straight lines

$a_1x + b_1y = c_1 = 0$

$a_2x + b_2y = c_2 = 0$

Homogenous Equation

An equation of the type

$a_0y^n + a_1y^{n-1}x + a_2y^{n-2}x^2 + ...a_nx^n = 0$ ...(i)

is called an homogenous equation of degree n, since the sum of powers of indices of x and y in every term is the same and equal to n.

Dividing equation (i) by x^n, we have

$$a_0\left(\frac{y}{x}\right)^n + a_1\left(\frac{y}{x}\right)^{n-1} + a_2\left(\frac{y}{x}\right)^{n-2} + ... + a_n = 0.$$

This equation being nth degree in y/x will have n roots. If m_1, m_2, m_3, ..., m_n are the roots of the above equation, then equation (i) can be expressed as

$$a_0\left(\frac{y}{x} - m_1\right)\left(\frac{y}{x} - m_2\right)\left(\frac{y}{x} - m_3\right)......\left(\frac{y}{x} - m_n\right) = 0$$

which implies that

$y - m_1x = 0$, $y - m_2x = 0$, $m_3x = 0$, ..., $m_nx = 0$

All these equation represent straight lines passing through the origin.

Thus a homogeneous equation of degree n represents n straight lines passing through the origin.

ANGLES BETWEEN THE LINES

To find the angle between the two straight lines represented by the equation

$$ax^2 + 2hxy + by^2 = 0$$

Proof:

Let the lines represented by this equation be

$$y - mx = 0$$

and $y - m'x = 0$

The joint equation of these two straight lines is given by

$$(y - mx)(y - m'x) = 0.$$

$$\therefore\ y^2 + \frac{2h}{b}xy + \frac{a}{b}x^2 \equiv (y - mx)(y - m'x)$$

Comparing the co-efficients of like terms, we have

$$m + m' = -2h/b$$

$$mm' = b/a$$

Let θ represent the angles between the straight lines, the we have

$$\tan\theta - \frac{m - m'}{1 + m_1m_2}$$

$$\frac{\sqrt{(m+m')^2-4mm'}}{1+m_1m_2}$$

$$\frac{\sqrt{\left(\frac{-2h}{b}\right)^2-4\frac{b}{a}}}{1+b/a}$$

$$\frac{\sqrt{4h^2-4ab}}{b+a}$$

$\therefore$ The angle between the lines $ax^2 + 2hxy + by^2 = 0$ is

$$\theta = \tan^{-1}\frac{2\sqrt{h^2-ab}}{b+a}$$

Corollary 1. *Condition for perpendicularity.*

The lines are perpendicular if $\theta = 90°$. The gives $a + b = 0$.

Corollary 2. *Condition for coincidence or parallel lines.*

The lines are coincidence if $\theta = 0$, i.e. if $h^2 = ab$

BISECTOR FORMULA

To find the equation of bisectors of the angle between the two straight lines represented by the equation

$$ax^2 + 2hxy + by^2 = 0.$$

Proof:

Let the equations of lines represented by $ax^2 + 2hxy + by = 0$ be

$y = mx$ and $y = m'x$. Then we have

$$\left.\begin{aligned} m+m' &= \frac{-2h}{b} \\ mm' &= a/b \end{aligned}\right\} \quad ...(i)$$

The equation of the bisectors of the angle between the lines are $y = mx$ and $y = m'x$ is given by

$$\frac{y-mx}{\sqrt{1+m^2}} = \pm\frac{y-m'x}{\sqrt{1+m'^2}}.$$

This gives the combined equation of the bisectors as

$$\left(\frac{y-mx}{\sqrt{1+m^2}}-\frac{y-m'x}{\sqrt{1+m'^2}}\right)\left(\frac{y-mx}{\sqrt{1+m^2}}+\frac{y-m'x}{\sqrt{1+m'^2}}\right)=0$$

$$\Rightarrow \qquad \frac{(y-mx)^2}{1+m^2}-\frac{(y-m'x)^2}{1+m'^2}=0.$$

$\Rightarrow y^2(m'^2 - m^2) + 2xy(m' - m)\ (1 - mm') + x^2(m^2 - m'^2) = 0$

$\Rightarrow y^2(m + m') + 2xy(1 - mm') + x^2(m + m') = 0$ (as $m_1 \neq m'_2$)

$$\Rightarrow y^2 + 2xy\left(\frac{1-mm'}{m+m'}\right)-x^2=0.$$

$$\Rightarrow y^2 - x^2 = -2xy\left\{\frac{1-a/b}{-2h/b}\right\}$$ [substituting $m + m'$ and mm' from (i)]

we have $y^2 - x^2 = \dfrac{b-a}{h}\,xy$

$$\Rightarrow \frac{x^2-y^2}{a-b}=\frac{xy}{h}$$

Hence the equation of the bisectors of the angle between the lines

$ax^2 + 2hxy + by^2 = 0$ is

$$\frac{x^2-y^2}{a-b}=\frac{xy}{h}\ \text{given by.}$$

CONDITION THAT A GENERAL EQUATION OF SECOND DEGREE REPRESENTS PAIR OF STRAIGHT LINES

The equation $ax^2 + 2hxy + by^2 + 2gx + 2fy + c = 0$ is called the general equation of second degree as it contains all the terms not only of second degree but also terms of lower degree. We now obtain the condition that general equation of second degree represents two straight lines.

Let $\quad ax^2 + 2hxy + by^2 + 2gx + 2fy + c = 0$ represents the two straight lines.

$$lx + my + n = 0 \quad \text{and} \quad l'x + m'y + n' = 0.$$

Then we have

$ax^2 + 2hxy + by^2 + 2gx + 2fy + c \equiv (lx + my + n)\ (l'x + m'y + n')$

Comparing the co-efficients of the like terms. We have

$$\frac{ll'}{a}=\frac{lm'+l'm}{2h}=\frac{mm'}{b}=\frac{ln'+l'n}{2g}=\frac{mn'+m'n}{2f}=\frac{nn'}{c}=\lambda(\text{say})$$

$\Rightarrow ll' = a\lambda$ (i); $lm' + l'm = 2h\lambda$ ...(iv)

$mm' = b\lambda$ (ii); $ln' + l'n = 2g\lambda$ ...(v)

$nn' = c\lambda$ (iii); $mn' + m'n = 2f\lambda$ ...(vi)

Multiplying the relations (iv), (v) and (vi), we have

$8fgh\ \lambda^3 = (lm' + l'm)(ln' + l'n)(mn' + m'n)$

$= 2\ ll'\ mm'\ nn' + ll'\ (m'^2n^2 + m^2n'^2)$
$+ mm'\ (n^2l'^2 + n'^2l^2) + nn'\ (l^2m'^2 + l'^2m^2)$

$= 2abc\ \lambda^3 + a\lambda\ (4f^2\lambda^2 - 2bc\ \lambda^2) + b\ \lambda\ (4g^2\lambda^2 - 2ac\ \lambda^2)$
$+ c\ \lambda\ (4h^2\lambda^2 - 2ab\ \lambda^2)$

$\Rightarrow 8fgh = -4abc + 4af^2 + 4bg^2 + 4ch^2$

$\Rightarrow abc + 2fgh - af^2 - bg^2 - ch^2 = 0.$

which is the required condition. That a general equation of second degree represent pair of straight this condition also be continue as

$$\begin{vmatrix} a & g & h \\ g & b & f \\ h & f & c \end{vmatrix}=0.$$

ANGLE BETWEEN THE LINES REPRESENTED BY THE EQUATION

$$ax^2 + 2hxy + by^2 + 2gx + 2fy + c = 0$$

Proof:

Let $l_1x + m_1y + n_1 = 0$ ...(i)

$l_2x + m_2y + n_2 = 0$ ...(ii)

be the lines represented by the given equation. we know that

$l_1l_2 = a\lambda,\ m_1m_2 = b\lambda,\ n_1n_2 = c\lambda,\ l_1m_2 + l_2m_1 = 2h\lambda,$

$m_1m_2 + m_2n_1 = 2f\lambda$ and $l_1n_2 + l_2n_1 = 2g\lambda,$

the slope of lines (i) and (ii) are $-l_1/m_1$ and $-l_2/m_2$.

Let θ be the angle between them.

$$\therefore\ \tan\theta=\frac{l_1/m_1-l_2/m_2}{l_1l_2\ m_1m_2}=\frac{l_1m_2-l_2m_1}{l_1l_2+m_1m_2}$$

$$= \frac{\sqrt{(l_1 m_2 + l_2 m_1)^2 - 4 l_1 l_2 m_1 m_2}}{l_1 l_2 + m_1 m_2}$$

$$= \frac{2\lambda\sqrt{h^2 - ab}}{\lambda(a+b)}$$

$$= \frac{2\sqrt{h^2 - ab}}{a+b}$$

Which is the required formula for calculating the angle between the two lines represented by the general equation of second degree. Condition for perpendicular is $a + b = 0$ condition for parallelogram is $h^2 = ab$

LINES JOINING THE ORIGIN TO THE INTERSECTION OF A CURVE AND A LINE

Let the equation of the curve be given by

$$ax^2 + 2hxy + by^2 + 2gx + 2fy + c = 0 \qquad \text{...(i)}$$

and the equ. of line is given as

$$lx + my = 1. \qquad \text{...(ii)}$$

Let the curve and the line intersect at the points A and B. Now the equation

$$ax^2 + 2hxy + by^2 + 2gx + 2hy + c\,(lx + my)^2 = 0 \qquad \text{...(iii)}$$

represents same curve passing through A and B. But the equation being homogenous in x and y of second degree represents pair of straight line passing through the origin and the points of inter section of a curve and a straight line. Thus the equation of the pairs of lines joining the origin and the points of intersection of a line and a curve is obtained by making the curve homogenous with the help of the line.

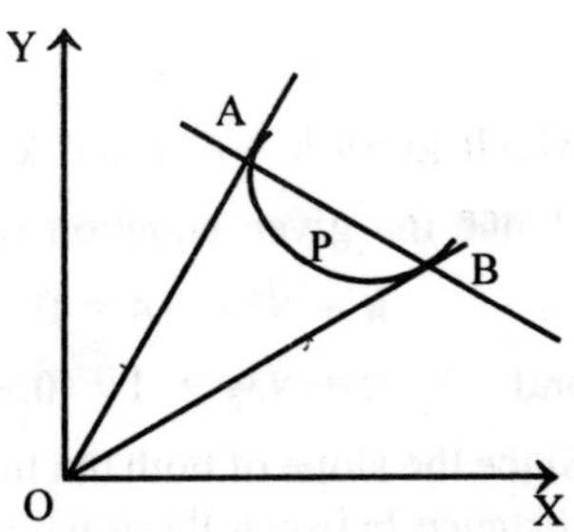

Example 1:

Show that the equation $x^2 + 2\sqrt{3}\,xy + 3y^2 - 3x - \sqrt{3}\,y - 4 = 0$ *represents two parallel straight lines and find the distance between them.*

Solution:

Comparing the given equation with general equation of second degree $ax^2 + 2hxy + by^2 + 2gx + 2fy + c = 0$, we have

$$a = 1,\ h = \sqrt{3},\ b = 3,\ g = -\frac{3}{2}\ f = -\frac{3}{2}\sqrt{3} \text{ and } c = -4.$$

Substituting these values in the required condition for pair of straight lines, i.e., in $abc + 2fgh - af^2 - bg^2 - ch^2$, we get

$$-12 + 2\left(-\frac{3}{2}\right)\sqrt{3}\,.\left(-\frac{3}{2}\right)\sqrt{3} - \frac{27}{4} - 3\left(\frac{9}{4}\right) + 4.3$$

$$= -12 + 2\,.\frac{27}{4} - \frac{27}{4} - \frac{27}{4} + 12$$

$= 0.$

This shows that the equation represents a pair of straight lines the given equation can be written as

$$(x + \sqrt{3}\ y)^2 - 3x - 3\sqrt{3}\ y - 4 = 0.$$

Let it represents the lines

$$x + \sqrt{3}\ y + k = 0$$

and $\quad x + \sqrt{3}\ y + k' = 0.$

This gives

$$(x + \sqrt{3}y)^2 - 3x - 3\sqrt{3}y - 4 \equiv (x + \sqrt{3}y + k)(x + \sqrt{3}y + k')$$

Comparing and equating coefficients of the like terms, we get

$$k + k' = -3$$

$$kk' = -4$$

which gives $k = -4$ and $k' = 1$.

Hence the given equation represents the straight lines

$$x + \sqrt{3}y - 4 = 0$$

and $\quad x + \sqrt{3}y + 1 = 0.$

Since the slope of both the lines is same, there fore, the lines are parallel. The distance between these parallel lines is the length of the prependicular distance from a point of one line to the other line. Clearly, $(-1, 0)$ lies on the line $x + \sqrt{3}y + 1 = 0$. Therefore, the perpendicular distance from $(-1, 0)$ to $x + \sqrt{3}y - 4 = 0$ is $\dfrac{-1-4}{\sqrt{(1+3)}} = \dfrac{-5}{2} = \dfrac{5}{2}$ (negative is dropped for distance)

Example 2:

Find the equation to the pair of straight lines joining the origin to the intersection of the straight line $y = mx + c$ and the curve $x^2 + y^2 = a^2$. Prove that they are at right angles if

$$2c^2 = a^2(1 + m^2)$$

Solution:

The equation of the curve is $x^2 + y^2 = a^2$ ...(i)

The equation of the straight line is $y = mx + c$ ...(ii)

Making (i) homogeneous with the help of (ii), we have

$$x^2 + y^2 = a^2\left(\frac{y - mx}{c}\right)^2$$

$$\Rightarrow x(c^2 - a^2m^2) = 2a^2mxy + y^2(c^2 - a^2) = 0. \quad ...(iii)$$

The equation (iii) being homogeneous will give two straight lines.

These lines will be at right angles if

coefficient of x^2 + coefficient of $y^2 = 0$.

$\Rightarrow c^2 - a^2m^2 + c^2 - a^2 = 0$

$\Rightarrow 2c^2 = a^2(1 + m^2)$

Hence proved.

MISCELLANEOUS EXAMPLES

Example 1:

Find the equations of the straight lines bisecting the angles between the pairs of straight lines given in,

(i) $4x^2 - 24xy + 11y^2 = 0$

(ii) $33x^2 - 71xy - 14y^2 = 0$

(iii) $x^2 - 2xy \sec\theta + y^2 = 0$

(iv) $x^2 + 2xy \cot\theta + y^2 = 0$

Solution:

(i) Equation of the bisectors of the angles between $4x^2 - 24xy + 11y^2 = 0$ will be

$$\frac{x^2 - y^2}{4 - 11} = \frac{xy}{-12};$$

$$12x^2 - 7xy - 12y^2 = 0.$$

(ii) Equation of the bisectors of the angles between

$33x^2 - 71xy - 14y^2 = 0$ is

$$\frac{x^2 - y^2}{33 + 14} = \frac{xy}{-\frac{71}{2}}$$

or $71x^2 + 94xy - 71y^2 = 0.$

(iii) Equations of the lines bisecting the angles between $x^2 + 2xy \sec\theta +$ $y^2 = 0$ will be $\dfrac{x^2 - y^2}{1 - 1} = \dfrac{xy}{\sec\theta}$ As R.H.S. is finite, denominator on L.H.S. is zero; hence the numerator of L.H.S. must be zero. Therefore the required equation is $x^2 - y^2 = 0$.

(iv) The equation of the bisectors of the angles between $x^2 + 2xy \cot\theta$ $+ y^2 = 0$ is $\dfrac{x^2 - y^2}{1 - 1} = \dfrac{xy}{\cot\theta}$

Hence as in part (3), the required equation is $x^2 - y^2 = 0$.

Example 2:

Find what straight lines are represented by the following equation and determine the angles between them

$$x^2 + 2xy \sec\theta + y^2 = 0.$$

Solution:

$x^2 + 2xy \sec\theta + y^2 = 0.$

$\Rightarrow m^2 + 2m \sec\theta + 1 = 0$ (dividing by x^2 and putting $y/x = m$)

$$\Rightarrow m = \frac{-2\sec\theta \pm \sqrt{(4\sec^2\theta - 4)}}{2} = -\sec\theta \pm \tan\theta = \frac{-1 \pm \sin\theta}{\cos\theta}$$

or $\dfrac{y}{x} = \dfrac{-1 \pm \sin\theta}{\cos\theta}$

Hence the equations are $y\cos\theta + x(1 + \sin\theta) = 0$

and $y\cos\theta + x(1 - \sin\theta) = 0.$

If ϕ be the angle between them, $\tan\phi = \dfrac{2\sqrt{(\sec^2\theta - 1)}}{1 + 1} = \tan\theta.$

$$\phi = \theta.$$

Example 3:

Find what straight lines are represented by the following equations and determine the angles between them

$$x^2 + 2xy \cot\theta + 1 = 0.$$

Solution:

(dividing by x^2 and putting $y/x = m$)

$$\Rightarrow \quad m = \frac{-2\cot\theta \pm \sqrt{(4\cot^2\theta - 4)}}{2} = \frac{-2\cot\theta \pm 2\sqrt{\left(\frac{\cos^2\theta}{\sin^2\theta} - 1\right)}}{2}$$

$$\Rightarrow \frac{y}{x} = -\frac{\cos\theta}{\sin\theta} \pm \sqrt{\left(\frac{\cos^2\theta - \sin^2\theta}{\sin^2\theta}\right)} = -\frac{\cos\theta}{\sin\theta} \pm \frac{\sqrt{(\cos 2\theta)}}{\sin\theta}$$

$$= -\frac{\cos\theta \pm \sqrt{(\cos 2\theta)}}{\sin\theta}$$

Hence the equations are $y\sin\theta + x\cos\theta - x\sqrt{(\cos 2\theta)} = 0$

and $\quad y\sin\theta + x\cos\theta + x\sqrt{(\cos 2\theta)} = 0.$

If ϕ be the angle between the lines, then

$$\tan\theta = \frac{2\sqrt{(\cot^2\theta - 1)}}{1+1} = \sqrt{\left(\frac{\cos^2\theta - \sin^2\theta}{\sin^2\theta}\right)} = \frac{\sqrt{(\cos 2\theta}}{\sin\theta}$$

$$= \operatorname{cosec}\theta\sqrt{(\cos 2\theta)}.$$

$$\therefore \quad \phi = \tan^{-1}[\operatorname{cosec}\theta\sqrt{(\cos 2\theta)}].$$

$$= 2\tan\alpha\sqrt{[(\sin^2\theta + \cos^2\theta\sin^2\alpha)\{\sin^2\theta + \cos^2\theta(1 - \cos^2\alpha)]}$$

$$= 2\tan\alpha\sqrt{\{(\sin^2\theta + \cos^2\theta\sin^2\alpha)\{\sin^2\theta + \cos^2\theta.\sin^2\alpha)\}}$$

$$= \tan\alpha(\sin^2\theta + \cos^2\theta\sin^2\alpha)$$

Denominator $= 2(\cos^2\theta\sin^2\theta + \sin^2\theta - \tan^2\alpha)$

$$= \frac{1}{\cos^2\alpha}\left[2\cos^2\theta\sin^2\alpha\cos^2\alpha + \sin^2\theta\cos^2\alpha - \sin^2\alpha\right]$$

$$= \frac{1}{\cos^2\alpha}\left[2\cos^2\theta\sin^2\alpha\cos^2\alpha + \sin^2\theta\cos^2\alpha - \cos^2\theta\sin^2\alpha - \sin^2\theta\sin^2\alpha\right]$$

Rearranging and factorizing, we get

$$= \frac{1}{\cos^2\alpha}\left[\sin^2\theta\left(\cos^2\alpha - \sin^2\alpha\right) + \cos^2\theta\,\sin^2\alpha\left(2\cos^2\alpha - 1\right)\right]$$

$$= \frac{\cos^2\alpha - \sin^2\alpha}{\cos^2\alpha}\left[\sin^2\theta + \cos^2\theta\,\sin^2\alpha\right]$$

$(\therefore\ 2\cos^2\alpha - 1 = \cos^2\alpha - \sin^2\alpha)$

$= (1 - \tan^2\alpha)\,[\sin^2\theta + \cos^2\theta\,\sin^2\alpha].$

Hence

$$\frac{\text{Numerator}}{\text{Deno min ator}} = \frac{2\tan\alpha\left(\sin^2\theta + \cos^2\theta\,\sin^2\alpha\right)}{\left(1 - \tan^2\alpha\right)\left(\sin^2\theta + \cos^2\theta\sin^2\alpha\right)} = \tan 2\alpha.$$

Therefore $\tan\phi = \tan 2\alpha$ by (1) or $\phi = 2\alpha$. **Proved.**

Example 4(a):

Prove that the two straight lines $(x^2 + y^2)(\cos^2\theta\sin^2\alpha + \sin^2\theta) = (x\tan\alpha - y\sin\theta)^2$ include an angle 2α.

Solution:

The equation of the lines is given as

$(x^2 + y^2)(\cos^2\theta\sin^2\theta + \sin^2\theta = (x\tan\alpha - y\sin\theta)^2$

$\Rightarrow x^2(\cos^2\theta\sin^2\alpha + \sin^2\theta - \tan^2\alpha)$

$+ y^2(\cos^2\theta\sin^2\alpha + \sin^2\theta - \sin^2\theta) + 2xy\tan\alpha\sin\theta = 0.$

$\Rightarrow x^2(\cos^2\theta\sin^2\alpha + \sin^2\theta - \tan^2\alpha)$

$+ y^2\cos^2\theta\sin^2\alpha + 2x\tan\alpha\sin\theta = 0.$

If ϕ be the angle between the lines represented by the equation.

$$\tan\phi = \frac{2\sqrt{\left[\left\{\tan^2\alpha\sin^2\theta - \left(\cos^2\theta\sin^2\alpha + \sin^2\theta - \tan^2\alpha\right)\cos^2\theta\sin^2\alpha\right\}\right]}}{\cos^2\theta\sin^2\alpha + \sin^2\theta - \tan^2\alpha + \cos^2\theta\sin^2\alpha\quad ...(1)}$$

Taking R.H.S.

Numerator

$$= 2\sqrt{\left\{\tan^2\alpha.\sin^2\theta - (\cos^2\theta\sin^2\alpha + \sin^2\theta - \tan^2\alpha).\cos^2\theta\sin^2\alpha\right\}}$$

$$2\sqrt{\left\{\tan^2\alpha\,(\sin^2\theta - \cos^4\theta\sin^2\theta\cos^2\alpha - \sin^2\theta\cos^2\theta\cos^2\alpha + \cos^2\theta\sin^2\right.}$$

$[\therefore\ \cos^2\theta\sin^2\alpha = \cos^2\theta\tan^2\alpha.\ \chi o\sigma^2\alpha]$

Rearranging and factorizing, we get

$$= 2\tan\ = 2\tan\alpha\sqrt{\{(\sin^2\theta + \cos^2\ \sin^2\alpha)\,(1 - \cos^2\theta\cos^2\alpha)\}}$$

$$= 2\tan\alpha\sqrt{\{(\sin^2\theta + \cos^2\ \sin^2\alpha)(\sin^2\theta + \cos^2\theta - \cos^2\theta\cos^2\alpha)\}}$$

$[\because\ 1 = \cos^2\theta + \sin^2\theta]$

Example 4(b):

Find what straight lines are represented by the following equation and determine the angle between them

$$x^2 - 7xy + 12y^2 = 0.$$

Solution:

The given equation is $x^2 - 7xy + 12y^2 = 0$. ... (1)

Factorizing, we get $(x - 3y)(x - 4y) = 0$.

Hence the equations represented by (1) are

$x - 3y = 0$ and $x - 4y = 0$.

If θ is the angle between the line then

$$\tan\theta = \frac{2\sqrt{\left(\frac{7}{2}\right)^2 - 1.12}}{1+12} = \frac{1}{13} \text{ (taking + ve sign)}$$

$$\Rightarrow \theta = \tan^{-1}\left(\frac{1}{13}\right).$$

Example 5(a):

Show that the two straight lines $x^2(\tan^2\theta + \cos^2\theta) - 2xy\tan\theta + y^2\sin^2\theta = 0$, make with the axis of x angles such that the difference of their tangents is 2.

Solution:

The given equation is

$$x^2(\tan^2\theta + \cos^2\theta) - 2xy\tan\theta + y^2\sin^2\theta = 0. \quad \text{... (1)}$$

Let the equation (1) represent two straight lines $y = m_1x$ and $y = m_2x$ where m_1 and m_2 are their slopes since (1) represents two straight lines passing through origin.

The combined equation will bc

$$(y - m_1x)(y - m_2x) = 0.$$

$$\Rightarrow \quad y^2 - (m_1 + m_2)xy + m_1m_2x^2 = 0 \quad \text{... (2)}$$

As (1) (2) represent same straight lines, the coefficients of x^2, y^2 and xy must be proportional i.e. $$\frac{1}{\sin^2\theta} = \frac{-(m_1+m_2)}{-2\tan\theta} = \frac{\tan^2\theta+\cos^2\theta}{m_1 m_2}$$

Hence $m_1 + m_2 = \dfrac{2\tan\theta}{\sin^2\theta} = \dfrac{2}{\sin\theta\cos\theta}$

Therefore $(m_1 - m_2)^2 = (m_1 + m_2)^2 - 4m_1m_2$

$$= \frac{4}{\sin^2\theta\cos^2\theta} - \frac{4\left(\sin^2\theta+\cos^4\theta\right)}{\sin^2\theta\cos^2\theta} = 4\left[\frac{1-\sin^2\theta-\cos^4\theta}{\sin^2\theta\cos^2\theta}\right]$$

$$= 4\left[\frac{\cos^2\theta-\cos^4\theta}{\sin^2\theta\cos^2\theta}\right] = \frac{4\cos^2\theta\left(1-\cos^2\theta\right)}{\cos^2\theta\sin^2\theta} = \frac{4\cos^2\theta\sin^2\theta}{\cos^2\theta\sin^2\theta}$$

Hence $(m_1 - m_2)^2 = 4$

or $m_1 - m_2 = \pm 2$

or $\tan\theta_1 \sim \tan\theta_2 = 2,$

where θ_1 and θ_2 are the inclinations of the lines represented by (1) from the x-axis.

Example 5(b):

Prove that the two straight lines $x^2\sin^2\alpha\cos^2\theta + 4xy\sin\alpha\sin\theta + y^2[4\cos\alpha - (1+\cos\alpha)^2\cos^2\theta] = 0$ meet at an angle α.

Solution:

The lines are given by

$x^2\sin^2\alpha\cos^2\theta + 4xy\sin\alpha\sin\theta + y^2$

$[4\cos\alpha - (1+\cos\alpha)^2\cos^2\theta] = 0.$

If ϕ be the angle between them, then we have

$$\tan\phi = \left\{\frac{2\sqrt{[(2\sin\alpha\sin\theta)^2 - \sin^2\alpha\cos^2\theta\,[4\cos\alpha-(1+\cos\alpha)^2\cos^2\theta)]}}{\sin^2\alpha\cos^2\theta + [4\cos\alpha-(1+\cos\alpha)^2\cos^2\theta]}\right\} \quad \ldots (1)$$

Taking R.H.S.

Numerator

$= 2\sqrt{[4\sin^2\alpha\sin^2\theta - (\sin^2\alpha\cos^2\theta)\{4\cos\alpha - (1+\cos\alpha)^2\cos^2\theta\}]}$

$= 2\sqrt{[\sin^2\alpha\{4\sin^2\theta - 4\cos\alpha\cos^2\theta + (1+\cos\alpha)^2\cos^4\theta\}]}$

$$= 2\sqrt{[\sin^2 \alpha\{4 - 4\cos^2\theta - 4\cos\alpha\cos^2\alpha\cos^2\theta + (1+\cos\alpha)^2\cos^4\theta\}]}$$

$$[\because \sin^2\theta = 1 - \cos^2\theta]$$

$$= 2\sqrt{[\sin^2 \alpha\{4 - 4\cos^2\theta\,(1+\cos\alpha) + (1+\cos\alpha)^2\cos^4\theta)]}$$

$$= 2\sqrt{[\sin^2 \alpha\{2 - (1+\cos\alpha)\cos^2\theta\}^2]}$$

$$= 2\sin\alpha\,[2 - (1+\cos\alpha)\cos^2\theta].$$

Denominator $= \sin^2\alpha\cos^2\theta + 4\cos\alpha - (1+\cos\alpha)^2\cos^2\theta.$

Rearranging and simplifying, we have

$$= 4\cos\alpha - (1 + \cos^2\alpha + 2\cos\alpha)\cos^2\theta + \sin^2\alpha\cos^2\theta$$

$$= 4\cos\alpha - \cos^2\theta\,[1 + \cos^2\alpha + 2\cos\alpha - \sin^2\theta].$$

$$= 4\cos\alpha - \cos^2\theta.[2\cos^2\alpha + 2\cos\alpha] \quad (\because 1 - \sin^2\alpha = \cos^2\alpha)$$

$$= 4\cos\alpha - \cos^2\theta.\ 2\cos\alpha\,(\cos\alpha + 1)$$

$$= 2\cos\alpha\,[2 - (1+\cos\alpha).\cos^2\theta].$$

$$\therefore \frac{\text{Numerator}}{\text{Denominator}} = \frac{2\sin\alpha\left[2-(1+\cos\alpha)\cos^2\theta\right]}{2\cos\alpha\left[2-(1+\cos\alpha)\cos^2\theta\right]} = \tan\alpha.$$

Hence (1), $\tan\phi = \tan\alpha$ or $\phi = \alpha$.

Example 5(c):

Prove that the following equation represents two straight lines; find also their point of intersection and the angle between them

$$6y^2 - xy - x^2 + 30y + 36 = 0.$$

Solution:

Here we have

$$6y^2 - xy - x^2 + 30y + 36 = 0.$$

Hence $a = -1$, $b = 6$, $c = 36$, $h = -1/2$, $g = 0$, $f = 15$,

$$\Delta = abc + 2fgh - af^2 - bg^2 - ch^2$$

$$= (-1).6.36 + 2.15.0\left(-\frac{1}{2}\right) - (-1)\,(15)^2 - 6.(0)^2 - 36\left(-\frac{1}{2}\right)^2 = 0$$

Hence the equation represents two straight lines.

Again $6y^2 - xy - x^2 = (3y + x)\,(2y - x)$

Hence let

$$6y^2 - xy - x^2 + 30y + 36 \equiv (3y + x + A) \times (2y - x + B). \quad \ldots(1)$$

Comparing coefficients of x and y, we get B – A = 0 ...(2)

3B + 2A = 30 ...(3)

and 2y – x + 6 = 0. ...(5)

If ϕ be the angle between them

$$\tan\theta = \frac{2\sqrt{\left[\left(-\frac{1}{2}\right)^2 - (-1)(6)\right]}}{-1+6} = \frac{2\sqrt{\left(\frac{25}{4}\right)}}{5} = 1.$$

$\therefore$ $\phi = 45^\circ$.

Solving (4) and (5), we get $x = \frac{6}{5}, y = -\frac{12}{5}$.

So point of intersection is $\left(\frac{6}{5}, -\frac{12}{5}\right)$.

Example 6:

Prove that the following equation represents two straight lines; find also their point of intersection and the angle between them

$$x^2 - 5xy + 4y^2 + x + 2y - 2 = 0.$$

Solution:

Here we have

$$x^2 - 5xy + 4y^2 + x + 2y - 2 = 0.$$

No we here a = 1, b = 4, c = –2, $h = -\frac{5}{2}, g = \frac{1}{2}$, f = 1

$$\therefore \Delta = 1.4(-2) + 2.1.\frac{1}{2}.-\left(\frac{5}{2}\right) - 1.(1)^2 - 4\left(\frac{1}{2}\right)^2 - (-2)\left(-\frac{5}{2}\right)^2 = 0.$$

Hence it represents two straight lines.

Again now $x^2 - 5xy + 4y^2 = (x - 4y)(x - y)$.

So let $x^2 - 5xy + 4y^2 + x + 2y - 2 \equiv (x - 4y + A)(x - y + B)$.

Comparing the coefficients of x and y, we get

A + B = 1

and –A–4B =2.

Solving, we get A = 2, B = –1.

Putting in (1), the required equations are

x – 4y + 2 = 0 ...(2)

and $\quad x - y - 1 = 0. \qquad \ldots (3)$

Solving (2) and (3), the point of intersection is (2, 1).

If θ is the angle between them, then we have

$$\tan\theta = \frac{2\left\{\sqrt{\left(-\frac{5}{2}\right)^2 - 1.4}\right\}^2}{1+4} = \frac{3}{5} \therefore \theta = \tan^{-1}\left(\frac{3}{5}\right).$$

Example 7:

Find the value of k so that the following equation may represent pairs of straight lines $12x^2 + xy - 6y^2 - 29x + 8y + k = 0$.

Solution:

The given equations $12x^2 + xy - 6y^2 - 29x + 8y + k = 0$

Here $a = 12$, $b = -6$, $c = k$, $f = 4$, $g = -\frac{29}{2}$, $h = \frac{1}{2}$

As it represents two straight lines, Δ must be zero.

Applying the condition $\Delta = 0$ we have

$$\Delta = 12(-6)k + 24\left(-\frac{29}{2}\right)^2\left(\frac{1}{2}\right) - 12(4)^2 - (-6)\left(-\frac{29}{2}\right)^2 - k\left(\frac{1}{2}\right)^2 = 0$$

$$\Rightarrow -72k - 58 - 192 + \frac{2523}{2} - \frac{k}{4} = 0.$$

Solving $k = 14$.

Example 8(a):

Find the value of k so that the following equation may represent pairs of straight lines

$$2x^2 + xy - y^2 + kx + 6y - 9 = 0.$$

Solution:

The given equation is

$$2x^2 + xy - y^2 + kx + 6y - 9 = 0$$

$$\Rightarrow \quad 4x^2 + 2xy - 2y^2 + 2kx + 12y - 18 = 0.$$

Here $a = 4$, $b = -2$, $c = -18$, $f = 6$, $g = k$, $h = 1$.

As it represents two straight lines, Δ must be zero.

$$\therefore \Delta = 4(-2)(-18) + 2(6)k.1 - 4(6)^2(-2)(k)^2 - (18)(1)^2 = 0$$

$\Rightarrow \quad 144 + 12k - 144 + 2k^2 + 18 = 0$

$\Rightarrow \quad k^2 + 6k + 8 = 0$ or $(k + 3)^2 = 0.$

Hence $\quad k = -3.$ **Ans.**

Example 8(b):

Prove that the equation $x^2 + 6xy + 9y^2 + 4x + 12y - 5 = 0$ represents two parallel lines.

Solution:

The given equation is

$x^2 + 6xy + 9y^2 + 4x + 12y - 5 = 0.$ ...(1)

Here a = 1, b = 9, c = –5, f = 6, g = 2, h = 3. Hence

$\Delta = 1.9\,(-5) + 2.6.2.3. -1.(6)^2 -9(2)^2 - (-5)\,(3)^2 = 0.$

So the equation represents 2 straight lines.

Again $\quad h^2 - ab = (3)^2 - 1.9 = 0$; so $h^2 = ab$.

Therefore, the lines represents by (1) are parallel.

Example 9(a):

Find the value of k so that following equation may represent pairs of straight lines $6x^2 + 11xy - 10y^2 + x + 31y + k = 0.$

Solution:

The given equation is

$6x^2 + 11xy - 10y^2 + x + 31y + k = 0.$

It will represent two straight lines if $\Delta = 0$.

Here $a = 6,\ b = -10,\ c = k,\ f = \frac{31}{2},\ f = \frac{1}{2},\ h = \frac{11}{2}$

$$\text{Now, } \Delta = 6\,(-10).k + 2.\frac{31}{2}, \frac{1}{2}, \frac{11}{2} - 6\left(\frac{31}{2}\right)^2 - (-10)\left(\frac{1}{2}\right)^2 - k.\left(\frac{11}{2}\right)^2 = 0.$$

$$\Rightarrow -60k + \frac{341}{2} - \frac{2883}{2} + \frac{5}{2} - \frac{121}{4}k = 0.$$

Solving, $k = -15.$ **Ans.**

Example 9(b):

Prove that the following equation represents two straight lines; find also their point of intersection and the angle between them

$3y^2 - 8xy - 3x^2 - 29x^2 - 29x + 3y - 18 = 0.$

Solution:

The given equation is

$$3y^2 - 8xy - 3x^2 - 29x + 3y - 18 = 0.$$

Here a = –3, b = 3, c = –18, $f = \frac{3}{2}, g = \frac{-29}{2}, h = -4$

Hence we have

$$\Delta = (-3)(3)(-18) + 2\left(\frac{3}{2}\right)\left(-\frac{29}{2}\right)(-4) - (5)\left(\frac{3}{2}\right)^2 - (3)\left(-\frac{29}{2}\right)^2 - (-18)(-4)^2 = 0.$$

Hence two straight lines are represented by

Now $3y^2 - 8xy - 3x^2 = (3y + x)(y - 3x)$. Hence let

$3y^2 - 8xy - 3x^2 - 29x + 3y - 18 \equiv (3y + x + A)(y - 3x + B)$ ...(1)

Equating the coefficients of x and y, we get

$$-3A + B = -29$$

and $$A + 3B = 3.$$

Solving, we get A = 9 and B = –2.

Substituting in (1), and equating each to zero, the equations are

$3y + x + 9 = 0$ ...(2)

and $y - 3x - 2 = 0.$ ...(3)

Solving (2) and (3), we get the point of intersection as $\left(-\frac{3}{2}, -\frac{5}{2}\right)$

If θ is the angle between them, then we have

$$\tan\theta = \frac{1\sqrt{\{(-4)^2 - (-3)(3)\}}}{-3+3} = \infty;\ \theta = 90^\circ$$

Example 10:

If the pairs of straight lines $x^2 - 2pxy - y^2$ 0 and $x^2 - 2qxy - y^2$ 0 be such that each pair bisects the angle between the other pair, prove that pq = – 1

Solution:

The given pairs of straight lines are

$x^2 - 2pxy - y^2 = 0$ ...(i)

$x^2 - 2qxy - y^2 = 0$ ...(ii)

Equation of the bisectors of angles of (i) are given by

$$\frac{x^2-y^2}{1-(-1)}=\frac{xy}{-p}\left(\text{using the equation } \frac{x^2-y^2}{a-b}=\frac{xy}{h}\right)$$

$$\Rightarrow x^2+\left(\frac{2}{p}\right) xy - y^2 = 0. \quad \text{...(iii)}$$

It is given that (ii) represents the equation of the bisectors of angles of (i). ∴ equation (ii) must represents the same pair of bisectors. This gives on comparison

$$\frac{1}{1}=\left(\frac{2/p}{-2q}\right)=\frac{1}{1} \text{ or } \frac{1}{p} = -q \text{ or } pq = -1,$$

which is the required result.

Example 11:

Prove that the general equation $ax^2 + 2hxy + by^2 + 2gx + 2fy + c = 0$ represents two parallel straight lines if $h^2 = ab$ and $bg^2 = af^2$. Also show that the distance between them is

$$2\sqrt{\frac{g^2-ac}{a(a+b)}}.$$

Solution:

The given equation is

$$ax^2 + 2hxy + by^2 + 2gx + 2fy + c = 0. \quad \text{...(i)}$$

Multiplying (i) by a we get

$$a^2x^2 + 2ax(hy + g) = -aby^2 - 2afy - ca$$

$$\text{or } (ax + hy + g)^2 = (hy + g)^2 - aby^2 - 2afy - ca$$

$$= (h^2 - ab)y^2 + 2(hg - af)y + g^2 - ca$$

$$(ax + hy + g) = \pm\sqrt{\left\{(h^2-ab)y^2+2(hg-af)y+(g^2-ca)\right\}} \quad \text{...(ii)}$$

If (i) represents two parallel straight lines, then the right hand side of (ii) must not contain any term involving y.

This gives

$$h^2 - ab = 0 \quad \text{...(iii)}$$

$$\text{and} \quad hg - af = 0 \quad \text{...(iv)}$$

(iv) can also be expressed as

$$h^2g^2 = a^2f^2$$

$$\Rightarrow \quad abg^2 = a^2f^2$$

$$\Rightarrow \quad bg^2 = af^2. \qquad ...(v)$$

Under these conditions equation (ii) reduces to

$$ax + hy + g = \pm\sqrt{(g^2 - ca)}$$

which gives two parallel lines as

$$ax + hy + g + \sqrt{g^2 - ca} = 0. \qquad ...(vi)$$

and

$$ax + hy + g - \sqrt{g^2 - ca} = 0. \qquad ...(vii)$$

Distance between these parallel lines is equal to the perpendicular distance from any point on one line to second line. Any point on line (vi) is $\left\{0, -\frac{(g + \sqrt{g^2 - ca})}{h}\right\}$

The perpendicular distance from this point on lines (vii) is

$$\frac{\left\{\frac{-(g + \sqrt{g^2 - ca})}{h}\right\} + g - \sqrt{(g^2 - ca)}}{\sqrt{(a^2 + h^2)}}$$

$$\Rightarrow \frac{-2\sqrt{g^2 - ca}}{\sqrt{(a^2 + h^2)}} = \frac{-2\sqrt{g^2 - c^2}}{\sqrt{(a^2 + ab)}}$$

Hence the require distance between these parallel line is

$$\frac{\sqrt{g^2 - ca}}{\sqrt{a(a + b)}}.$$

Example 12:

If the lines $ax^2 + 2hxy + by^2 = 0$ *be the sides of a parallelogram and the line* $lx + my = 1$ *be one of its diagonals, show that the equation of the other diagonal is* $y\ (bl - hm) = x\ (am - hl)$

Solution:

Since the other diagonal will pass through the origin, let its equation be

$y = m'x.$ ...(i)

The point of intersection of these two diagonals is

$$\left(\frac{1}{l+mm'}, \frac{m'}{l+mm'}\right).$$

The point of intersection of the line $lx + my = 1$

and $ax^2 + 2hxy + by^2 = 0$ are given by

$$ax^2 + 2hx\left(\frac{1-lx}{m}\right) + b\left(\frac{1-lx}{m}\right)^2 = 0$$

$\Rightarrow (am^2 - 2hm + bl^2)\, x^2 + 2(hm - bl)\, x + b = 0.$

This gives

$$x_2 + x_2 = -\frac{2(hm-bl)}{am^2-2hlm+bl^2}, \qquad \text{...(ii)}$$

where x_1 and x_2 are the abscissas of the points of intersection of the curve $ax^2 + 2hxy + by^2$ and $lx + my = 1$.

Since the diagonals of parallelogram bisect each other, it gives

$$\frac{1}{l+mm'} = \frac{bl-hm}{am^2-2hlm+bl^2}$$

$$\therefore\ m' = \frac{am-hl}{bl-hm}.$$

Substituting m' in (i), the equation of other diagonal becomes

$$y = \frac{am-hl}{bl-hm}x$$

or $y(bl - hm) = x(am - hl)$,

which is the required result.

Example 13:

Prove that the following equation represents two straight lines; find also their point of intersection and the angle between them

$y^2 + xy - 2x^2 - 5x - y - 2 = 0.$

Solution:

The given equation is

$$y^2 + xy - 2x^2 - 5x - y - 2 = 0.$$

Here $a = -2$, $b = 1$, $c = -2$, $f = -\frac{1}{2}, g = -\frac{5}{2}, h = \frac{1}{2}.$

Hence

$$\Delta = (-2)\ (1)\ (-2) + 2.$$

$$\left(-\frac{12}{2}\right)\left(-\frac{5}{2}\right)\left(\frac{1}{2}\right) - (-2)\left(-\frac{1}{2}\right)^2 - 1.\left(-\frac{5}{2}\right)^2 - (-2).\left(\frac{1}{2}\right)^2 = 0.$$

So the equation represents two straight lines are given as

Now, $y^2 + xy - 2x^2 = (y - x)\ (y + 2x)$. Hence let

$$y^2 + xy - 2x^2 - 5x - y - 2 \equiv (y - x + A)\ (y + 2x + B). \qquad ...(1)$$

Comparing the coefficients of y and x, we get

$$A + B = -1 \text{ and } 2A - B = -5.$$

Solving, we get $A = -2$, $B = 1$.

Substituting in (1) and putting each factor equal to zero, the equations are

$$y - x - 2 = 0 \qquad ...(2)$$

and $$y + 2x + 1 = 0. \qquad ....(3)$$

Solving (2) and (3), the point of intersection is $(-1, 1)$.

If θ be the angle between them, then we have

$$\tan\theta = \frac{2\sqrt{\{(1/2)^2 - (-2)\ (1)\}}}{-2-1} = 3 \text{ (omitting –ve sign)}$$

or $$\theta = \tan^{-1} 3.$$

Important Condition. If the two lines represented by $ax^2 + 2hxy + by^2 + 2gx + 2fy + c = 0$, be perpendicular to each other, then $a + b = 0$ and if, they are parallel, then $h^2 = ab$ and conversely, if $a + b = 0$, the lines are perpendicular and if $h^2 = ab$, the lines are parallel.

Example 14:

Find the value of k so that the following equation may present pairs of straight lines $12x^2 - 10xy + 2y^2 + 11x - 5y + k = 0$.

Solution:

The given equation is

$$12x^2 - 10xy + 2y^2 + 11x - 5y + k = 0.$$

Here $a = 12, b = 2, c = k, f = -\frac{5}{2}, g = \frac{11}{2}, h = -5.$

If it represents 2 straight lines, $\Delta = 0$.

$$\Delta = 12.2k + 2.\left(-\frac{5}{2}\right)\left(\frac{11}{2}\right)(-5) - 12\left(-\frac{5}{2}\right)^2 - 2\left(\frac{11}{2}\right)^2 - k(-5)^2 = 0.$$

$\Rightarrow 24k + \frac{275}{2} - 75 - \frac{121}{2} - 25k = 0.$ Solving $k = 2$. **Ans.**

Example 15(a):

Find the value of k so that the following equation may represent pairs of straight lines $12x^2 + kxy + 2y^2 + 11x - 5y + 2 = 0$.

Solution:

The given equation is

$$12x^2 + kxy + 2y^2 + 11x - 5y + 2 = 0.$$

$$\Rightarrow \quad 24x^2 + 2kxy + 4y^2 + 22x - 10y + 4 = 0.$$

Now $a = 24, b = 4, c = 4, f = -5, g = 11, h = k.$

If it represents 2 straight lines, $\Delta = 0$.

$$\Delta = 24.4.4 + 2.(-5)\,11.k - 24.(-5)^2 - 4\,(11)^2 - 4k^2 = 0.$$

$$\Rightarrow \quad -4k^2 - 110k - 700 = 0$$

$$\Rightarrow \quad 2(k + 10)(2k + 35) = 0$$

Hence either $k + 10$ or $2x + 35 = 0$

$$\therefore\ k = -10 \text{ or } k = -\frac{35}{2}.$$

Example 15(b):

Find the value of k so that the following equation may represent pairs of straight lines $x^2 + kxy + y^2 - 5x - 7y + 6 = 0$.

Solution:

The given equation is

$$x^2 + kxy + y^2 - 5x - 7y + 6 = 0.$$

$$\Rightarrow 2x^2 + 2kxy + 2y^2 - 10x - 14y + 12 = 0.$$

Here $a = 2, b = 2, c = 12, f = -7, g -5, h = k.$

If it represents 2 straight lines, $\Delta = 0$.

$$\therefore \Delta = 2.2.12 + 2\,(-7)\,(-5)\,k - 2\,(-7)^2 - (-5)^2 - 12k^2 = 0$$

$\Rightarrow \quad -12k^2 + 70k - 100 = 0$

or $\quad 6k^2 - 35k + 50 = 0$

$\Rightarrow \quad (3k - 10)(2k - 5) = 0$

or $\quad k = 10/3$ or $5/2$.

$k = 5/2$ **Ans.**

Example 15(c):

Prove that the equations to the straight lines passing through the origin which make an angle α with the straight line y + x = 0 are given by the equation $x^2 + 2xy \sec 2\alpha + y^2 = 0$.

Solution:

Equation of any line passing through origin is y = mx. ...(1)

The given equation is y + x = 0 ...(2)

If α is the angle between (1) and (2), then $\tan\alpha = \dfrac{-1-m}{-1+m}$

[∴ slope of (2) is –1]

$$\Rightarrow m = \frac{\tan\alpha - 1}{\tan\alpha + 1} = -m_1 \text{ (say).} \quad ...(3)$$

Again, if the angle between (1) and (2) be – α, then $-\tan\alpha = \dfrac{-1-m}{-1+m}$

$$\text{or } m = \frac{\tan\alpha + 1}{\tan\alpha - 1} = m_2 \text{ (say).} \quad ...(4)$$

Putting the values of m_1 and m_2 from (3) and (4) in (1), the two equations are

$$\left(y - \frac{\tan\alpha - 1}{\tan\alpha + 1}\right) = 0 \text{ and } \left(y - \frac{\tan\alpha + 1}{\tan\alpha - 1}x\right) = 0$$

So combined equation is $\left(y - \dfrac{\tan\alpha - 1}{\tan\alpha + 1}x\right)\left(y - \dfrac{\tan\alpha + 1}{\tan\alpha - 1}x\right) = 0$

$\Rightarrow y^2 + 2xy \sec 2\alpha + x^2 = 0$

$$\Rightarrow \left[\because -\left(\frac{\tan\alpha - 1}{\tan\alpha + 1} + \frac{\tan\alpha + 1}{\tan\alpha - 1}\right) = 2\sec 2\alpha\right].$$

Example 15(d):

Find the value of k so that the following equation may represent Paris of straight lines

$$6x^2 + xy + ky^2 - 11x + 43y - 35 = 0.$$

Solution:

The given equation is

$6x^2 + xy + ky^2 - 11x + 43y - 35 = 0.$

$\Rightarrow 12y^2 + 2xy + 2ky^2 - 22x + 86y = 70 = 0$ (multiplying by 2).

If it represents two straight lines, Δ must be zero.

Here $a = 12, b = 2k, c = -70, h = 1.\ a = 43,\ g = -11$

$\therefore \Delta = 12.2k.(-70) + 2.43, (-11)(1) - 12(43)^2 - 2k(-11)^2 - (-70)(1)^2 = 0.$

or $-1680k - 946 - 22188 - 242k + 70 = 0.$

Solving, we get $k = -12$. **Ans.**

Example 16(a):

Find the value of k so that the following equation may represent pairs of straight lines kxy – 8x + 9y – 12 = 0.

Solution:

The given equation is

$$kxy - 8x + 9y - 12 = 0$$

or $$2kxy - 16x + 18y - 24 = 0.$$

Here $a = 0, b = 0, c = -24, f = 9, g = -8, h = k.$

As it represents 2 straight lines, $\Delta = 0$. Then we have

$\Delta = 0.0\,(-24) + 2.9.(-8)\,(k) - 0\,(9)^2 - 0(-8)^2 - (-24)\,k^2 = 0$

or $$24k^2 - 144k = 0.$$

Solving, we get $k = 0$ or $k = 6$. **Ans.**

If $k = 0$, it will give an equation of 1st degree. Hence

$k = 6.$ **Ans.**

Example 16(b):

Find the value of k so that the following equation may represent pairs of straight lines $x^2 + 10/3xy + y^2 - 5x - 7y + k = 0$.

Solution:

The given equation is

$$x^2 + \frac{10}{3}xy + y^2 - 5x - 7y + k = 0.$$

Here we have

$$a = 1, b = 1, c = k,\ f = -\frac{7}{2}, g = -\frac{5}{2}, h = \frac{5}{3}.$$

As it represents 2 straight lines, $\Delta = 0$.

$$\therefore \Delta = 1.1.k + 2\left(-\frac{7}{2}\right)\left(-\frac{5}{2}\right)\left(\frac{5}{3}\right) - 1.\left(-\frac{7}{2}\right)^2 - 1\left(\frac{5}{2}\right)^2 - k\left(\frac{5}{3}\right)^2 = 0$$

$$\Rightarrow k + \frac{175}{6} - \frac{49}{4} - \frac{25}{9}k = 0.$$

Solving $k = 6$. **Ans.**

Example 17:

Prove that the equation y^2 (cos α + $\sqrt{3}$ sin α) cos α – xy (sin 2α – $\sqrt{3}$ cos 2α) + x^2 (sin α – $\sqrt{3}$ cos α) sin α = 0, represents two straight lines inclined at 60º to each other.

Prove also that the area of the triangle formed with them by the straight line (cos α – $\sqrt{3}$ sin α) y – (sin α + $\sqrt{3}$ cos α) x + α = 0 is a2|4 $\sqrt{3}$, and that this triangle equilateral.

Solution:

The given equation is

$y^2 (\cos\alpha + \sqrt{3}\sin\alpha)\cos\alpha - xy(\sin 2\alpha - \sqrt{3}\cos 2\alpha) + x^2(\sin\alpha - \sqrt{3}\cos\alpha)\sin\alpha = 0$.

As the given equation is a homogenous equation in second degree, it represents 2 straight lines passing through origin.

As $-(\sin 2\alpha - \sqrt{3}\cos 2\alpha) = (\sqrt{3}\cos 2\alpha - \sin 2\alpha)$

$= (\sqrt{3}\cos^2\alpha - \sqrt{3}\sin^2\alpha - 2\sin\alpha\cos\alpha)$

$= (\sqrt{3}\cos\alpha + \sin\alpha)(\cos\alpha - \sqrt{3}\sin\alpha)$.

The given equation can be factorized as

$(y\cos\alpha - x\sin\alpha)\{y(\cos\alpha + \sqrt{3}\sin\alpha) - x(\sin\alpha - \sqrt{3}\cos\alpha)\} = 0$.

Hence the given equation represents $y\cos\alpha - x\sin\alpha = 0$ ...(1)

and $y(\cos\alpha + \sqrt{3}\sin\alpha) - x(\sin\alpha - \sqrt{3}\cos\alpha) = 0$.

If θ be the angle between them, then we have

$$\tan\theta = \pm\frac{\dfrac{\sin\alpha - \sqrt{3}\cos\alpha}{\cos\alpha + \sqrt{3}\sin\alpha} - \dfrac{\sin\alpha}{\cos\alpha}}{1 + \dfrac{\sin\alpha - \sqrt{3}\sin\alpha}{\cos\alpha + \sqrt{3}\cos\alpha}\cdot\dfrac{\sin\alpha}{\cos\alpha}} = \pm\sqrt{3} \quad \text{(on simplification).}$$

$$1+\frac{\sin\alpha-\sqrt{3}\cos\alpha}{\cos\alpha+\sqrt{3}\sin\alpha}\cdot\frac{\sin\alpha}{\cos\alpha}$$

$\therefore \theta = 60°$.

The third given straight line is

$(\cos\alpha - \sqrt{3}\sin\alpha)\,y - (\sin\alpha + \sqrt{3}\cos\alpha)\,x + \alpha = 0.$ ...(3)

The point of intersection of (1) and (2) say A, is clearly $\equiv (0, 0)$.

The point of intersection of (1) and (3) say B, will be

$$\equiv\left(\frac{a\cos\alpha}{\sqrt{3}}, \frac{a\sin\alpha}{\sqrt{3}}\right).$$

If ϕ be the angle between (1) and (3), then

$$\tan\phi = \pm\frac{\dfrac{\sin\alpha}{\cos\alpha}-\dfrac{\sin\alpha+\sqrt{3}\cos\alpha}{\cos\alpha-\sqrt{3}\sin\alpha}}{1+\dfrac{\sin\alpha}{\cos\alpha}\cdot\dfrac{\sin\alpha+\sqrt{3}\cos\alpha}{\cos\alpha-\sqrt{3}\sin\alpha}} = \pm\sqrt{3}.$$

$\therefore \qquad \phi = 60°$.

So two of ΔABC are 60° each; hence third will also be 60°. Therefore the triangle is equilateral triangle.

Area of an equilateral triangle with side x is $= \frac{x^2\sqrt{3}}{4}$.

The side AB of the triangle ABC is

$$= \sqrt{\left\{\left(\frac{a\cos\alpha}{\sqrt{3}}-0\right)^2+\left(\frac{a\sin\alpha}{\sqrt{3}}-0\right)^2\right\}} = \pm\sqrt{3}.$$

The area of the given triangle is $= \left(\frac{a}{\sqrt{3}}\right)^2\cdot\frac{\sqrt{3}}{4} = \frac{a^2}{4\sqrt{3}}$.

Example 18:

Find what straight lines are represented by the following equation and determine the angles between them $y^2 - 16 = 0$.

Solution:

$y^2 - 16 = 0$ or $(y - 4)(y + 4) = 0$.

Hence the equations are $y = 4$ or $y = -4$.

Clearly the lines are parallel to x-axis.

Example 19:

Find what straight lines are represented by the following equation and determine the angle between them $y^3 - xy^2 - 14x^2y + 24x^3 = 0$.

Solution:

$$y^3 - xy^2 - 14x^2y + 24x^3 = 0.$$

Dividing the equation by x^3 and putting $y/x = m$, we get

$$m^3 - m^2 - 14m + 24 = 0.$$

Factorizing by remainder theorem, we get

$$(m - 2)(m - 3)(m + 4) = 0.$$

Putting the value of m, $\left(\frac{y}{x}-2\right)\left(\frac{y}{x}-3\right)\left(\frac{y}{x}+4\right) = 0.$

$$(y - 2x)(y - 3x)(y + 4x) = 0.$$

Hence the lines represented will be

$$y - 2x = 0,\ y - 3x = 0,\ y + 4x = 0.$$

If θ_1 be the angle between $y - 2x = 0$ and $y - 3x = 0$, then

$$\tan\theta_1 = \frac{3-2}{1+3.2} = \frac{1}{7} \quad \left(\therefore\ \tan\theta = \frac{m_1 - m_2}{1+m_1 m_2}\right)$$

or $\theta_1 = \tan^{-1}\left(\frac{1}{7}\right).$

Similarly, if θ_2 be the angle between $y - 2x = 0$ and $y + 4x = 0$.

$$\tan\theta_2 = \frac{2+4}{1+(2)(-4)} = \left(\frac{-6}{7}\right) \quad \therefore\ \theta_2 = \tan^{-1}\left(\frac{-6}{7}\right).$$

If θ_3 angle between $y - 3x = 0$ and $y + 4x = 0$, then

$$\tan\theta_3 = \frac{4+3}{1+4(-3)} = \frac{7}{-11}$$

$$\theta_3 = \tan^{-1}\left(\frac{-7}{13}\right)$$

Example 20(a):

Find what straight lines are represented by the following equations and determine the angle between them $4x^2\ 24\ x + 11y^2 = 0$.

Solution:

$$4x^2 - 24\,xy + 11y^2 = 0$$

Factorizing, we get $(2x - y)\,(2x - 11y) = 0.$

Hence equations are $2x - y = 0$ and $2x - 11y = 0.$

$$\tan\theta = \frac{2\sqrt{\{(-12)^2 - 4.11\}}}{4+11} = \frac{20}{15}$$

or $$\theta = \tan^{-1}\left(\frac{4}{3}\right).$$

Example 20(b):

Find what straight lines are represented by the following equation and determine the angle between them

$$33x^2 - 71xy - 14y^2 = 0.$$

Solution:

$$33x^2 - 71y - 14y^2 = 0.$$

Factorizing, $(11x + 2y)\,(3x - 7y) = 0.$

Hence the equations are $11x + 2y = 0$

and $3x - 7y = 0.$

$$\tan\theta = \frac{2 - \sqrt{\left\{\left(-\frac{71}{2}\right)^2 + 33.14\right\}}}{33-14} = \frac{83}{19} \quad \therefore\ \theta = \tan^{-1}\left(\frac{83}{19}\right).$$

Example 20(c):

Find what straight lines are represented by the following equation and determine the angles between them

$$x^3 - 6x^2 - 11x - 6 = 0.$$

Solution:

By remainder theorem, the factors are

$$(x - 1)(x - 2)(x - 3) = 0.$$

Hence, the lines represented will be

$$x - 1 = 0,$$

$$x - 2 = 0$$

and $$x - 3 = 0,$$

and clearly all the lines are parallel to y-axis. **Ans.**

Example 21:

Prove that the angle between the straight lines joining the origin to the intersection of the straight line y = 3x + 2 with the curve

$$x2 + 2xy + 3y^2 + 4x + 8y - 11 = 0 \text{ is } \tan^{-1}\frac{2\sqrt{2}}{3}.$$

Solution:

The equation of the curve is

$$x^2 + 2xy + 3y^2 + 4x + 8y - 11 = 0$$

The line is given as

$$y = 3x + 2 \text{ or } \frac{y}{2} - \frac{3x}{2} = 1. \qquad ..(1)$$

The curve is given as $x^2 + 2xy + 3y^2 + 4x + 8y - 11 = 0$ ...(2)

Making (2) homogeneous with the help of (1), we get

$$x^2 + 2xy + 3y^2 + 4x\left(\frac{y}{2} - \frac{3x}{2}\right) + 8y\left(\frac{y}{2} - \frac{3x}{2}\right) - 11\left(\frac{y}{2} - \frac{3x}{2}\right)^2 = 0$$

$$\Rightarrow x^2\left(1 - 6 - \frac{99}{4}\right) + y^2\left(3 + 4 - \frac{11}{4}\right) + xy\left(2 + 2 - 12 + \frac{33}{2}\right) = 0$$

$$\Rightarrow \quad -\frac{119}{4}x^2 + \frac{17}{4}y^2 + \frac{17}{2}xy = 0$$

$$\Rightarrow \quad 7x^2 - 2xy - y^2 = 0.$$

If ϕ be the angle between them, then

$$\tan\theta = \frac{\pm 2\sqrt{\{(-1)^2 - 7(-1)\}}}{7 + (-1)} = \frac{2\sqrt{8}}{6} = \frac{2\sqrt{2}}{3}.$$

$$\therefore \theta = \tan^{-1}\left(\frac{2\sqrt{2}}{3}\right).$$

Example 22:

What loci are represented by the equation xy – ay = 0.

Solution:

We have $xy - ay = 0 \quad \Rightarrow y(x - a) = 0$

So the given line represents 2 straight lines as

$y = 0$, i.e., x-axis and $(x - a) = 0$,

i.e., parallel to y-axis at a distance a.

Example 23(a):

What loci are represented by the equation $x^3 - x^2 - x + 1 = 0$.

Solution:

The given equation is

$$x^3 - x^2 - x + 1 = 0$$

$\Rightarrow$ $(x - 1)(x - 1)(x + 1) = 0$ (on factorizing).

Hence the equation represents 3 straight lines, given by

$x - 1 = 0$, $x - 1 = 0$ and $x + 1 = 0$.

the first two coinciding and all are parallel to y-axis at a distance $\pm$ 1.

Example 23(b):

What loci are represented by the equation $x^3 - xy^2 = 0$.

Solution:

The given equation is

$x^3 - xy^2 = 0$ or $x(x - y)(x + y) = 0$.

Hence the equation represents 3 straight lines as

$x = 0$, i.e., y-axis $x - y = 0$ and $x + y = 0$,

last two being inclined at an angle of 45° and 135° to x-axis and passing through origin.

Example 24:

Show that the straight lines joining the origin to the other two points of intersection of the curves whose equations are $ax^2 + 2hxy + by^2 + 2gx = 0$ *and* $a'x^2 + 2h'xy + b'y^2 + 2g'x = 0$ *will be at right angles if* $g(a' + b') - g'(a + b) = 0$.

Solution:

The two curves are given by

$$ax^2 + 2hxy + by^2 + 2gx = 0, \quad ...(1)$$

$$a'x^2 + 2h'xy + b'y^2 + 2g'x = 0. \quad ...(2)$$

To get the combined equation of the straight lines joining the point of intersection of (1) and (2) to the origin, we make (2) homogeneous with the help of (1), i.e., we eliminate the terms in first degree from (1) and (2). Hence multiplying (1) by g' and (2) by g and subtracting, we get

$(g'ax^2 + 2g'hxy + g'by^2 + 2gg'x) - (ga'x^2 + 2h'gxy + gb'y^2 + 2gg'x) = 0$

$\Rightarrow x^2 (ag' - ga') + 2 (g'h - gh') xy + (g'b - gb') y^2 = 0,$ ...(3)

which is the required equation.

If the lines represented by (3) be at right angles, then

$$(ag' - ga') + (g'b - gb') = 0$$

$$\Rightarrow \quad g' (a + b) - g (a' + b') = 0$$

$$\Rightarrow \quad g (a' + b') - g' (a + b) = 0.$$

Example 25:

Show that the straight lines $(A^2 - 3B^2) x^2 + 8ABxy + (B^2 - 3A^2) y^2 = 0$, form with the line $Ax + By + C = 0$ an equilateral triangle whose area is $C^2/\sqrt{3}(A^2 + B^2)$.

Solution:

The given equation is

$$(A^2 - 3B^2)(x^2 + 8ABxy + (B^2 - 3A^2) y^2 = 0.$$

As it is homogenous equation in second degree, so represents 2 straight lines passing through origin.

Factorizing, we get

$\{(\sqrt{3}A - B) y - (A + \sqrt{3}B)x\} \{(\sqrt{3}A + B) y + (A - \sqrt{3}B) x\} = 0.$

Hence the lines represented by the given equation will be

$$(\sqrt{3}A - B) y - (A + \sqrt{3}B) x = 0 \quad ...(1)$$

and $$(\sqrt{3}A + B) y + (A - \sqrt{3}B) x = 0 \quad ...(2)$$

The third line is given as

$$Ax + By + C = 0. \quad ...(3)$$

If α be the angle between (1) and (3), β be the angle between (2) and (3), then

$$\tan\alpha = \pm\left\{\frac{\left(\frac{A+\sqrt{3}B}{\sqrt{3}A-B}\right) - \left(-\frac{A}{B}\right)}{1+\left(\frac{A+\sqrt{3}B}{\sqrt{3}A-B}\right) - \left(-\frac{A}{B}\right)}\right\} = \pm\frac{\sqrt{3}\left(A^2+B^2\right)}{A^2+B^2} = \sqrt{3}$$

$\therefore \quad \alpha = 60^\circ$ (taking acute angle).

Similarly,

$$\tan\beta = \pm\left\{\frac{\left(\frac{A-\sqrt{3}B}{\sqrt{3}A+B}\right)-\left(-\frac{A}{B}\right)}{1+\left(\frac{A-\sqrt{3}B}{\sqrt{3}A+B}\right)-\left(\frac{A}{B}\right)}\right\} = \pm\frac{\sqrt{3}\left(A^2+B^2\right)}{A^2+B^2} = \pm\sqrt{3}$$

$\therefore \quad \beta = 60^\circ.$

As two angles of the triangle are 60° each, the triangle is an equilateral one.

If p be the altitude of an equilateral triangle, its area may be given by

$$\text{Area} = \frac{p^2}{\sqrt{3}}.$$

The perpendicular distance from origin (vertex of the Δ) on opposite side

$Ax + By + C = 0$ is

$$\frac{C}{\sqrt{(A^2+B^2)}}.$$

Hence the area of the triangle $= \dfrac{C^2}{\sqrt{3}\left(A^2+B^2\right)}$. unit **Ans.**

Example 26:

Prove that the equation $y^3 - x^3 + 3xy(y-x) = 0$, represents three straight lines equally inclined to one another.

Solution:

The given equation is

$$y^3 - x^3 + 3xy(y-x) = 0$$

$\Rightarrow \quad (y-x)(y^2 + xy + x^2) + 3xy(y-x) = 0.$

$\Rightarrow \quad (y-x)(y^2 + xy + x^2 + 3xy) = 0.$

$\Rightarrow \quad (y-x)[y + (2-\sqrt{(3)}\,x](y + (2+\sqrt{(3)}\,x] = 0.$

Hence $\quad y - x = 0 \qquad$ ...(1)

$y + (2-\sqrt{3})x = 0 \qquad$ ...(2)

and $\quad y + (2+\sqrt{3})x = 0 \qquad$ ...(3)

will be the three lines represented by the given equations.

If α_1, α_2 and α_3 be angles between (1), (2); (2), (3) and (1), (3), respectively, then we have

$$\tan\alpha_1 = \pm\frac{1+(2-\sqrt{3})}{1-1(2-\sqrt{3})} = \pm\frac{3-\sqrt{3}}{\sqrt{3}-1} = \pm\frac{\sqrt{3}(\sqrt{3}-1)}{\sqrt{3}-1} = \pm\sqrt{3},$$

$$\tan \alpha_2 = \pm \frac{-(2-\sqrt{3})+(2+\sqrt{3})}{1+(2-\sqrt{3})\,(2+\sqrt{3})} = \pm \frac{2\sqrt{3}}{2} = \pm\sqrt{3},$$

$$\tan \alpha_2 = \pm \frac{1+(2+\sqrt{3})}{1-(2+\sqrt{3})} = \pm \frac{3+\sqrt{3}}{-\sqrt{3}-1} = \pm \frac{3\,(\sqrt{3}+1)}{\sqrt{(3+1)}} = \pm\sqrt{3}.$$

Hence $\alpha_1 = \alpha_2 = \alpha_3$.

Example 27:

The equations to a pair of opposite sides of a parallelogram are $x^2 - 7x + 6 = 0$ and $y^2 - 14y + 40 = 0$; find the equations to its diagonals.

Solution:

The equations of one set of opposite sides is $x^2 - 7x + 6 = 0$ or $(x - 6)(x - 1) = 0$.

Hence the equations of the opposite sides say AB and CD respectively are

$$x - 6 = 2 \qquad ...(1)$$

and $$x - 1 = 0 \qquad ...(2)$$

The equation of the other set of opposite sides is

$$y^2 - 14y + 40 = 0$$

or $$(y - 4)(y - 4)(y - 10) = 0.$$

So the equations of other two sides say AD and BC respectively are

$$y - 4 = 0 \qquad ...(3)$$

and $$y - 10 = 0. \qquad ...(4)$$

Solving (1) and (3), the co-ordinates of A are (6, 4).

Solving (1) and (4), the co-ordinates of B are (6, 10).

Solving (2) and (4), the co-ordinates of C are (1, 10).

Solving (2) and (3), the co-ordinates of D are (1, 4).

Equation of AC is $y - 4 = \frac{10-4}{1-6}(x-6)$

$\Rightarrow$ $5y + 6x = 56.$

Equation of BD is $y - 10 = \frac{4-10}{1-6}(x-6)$

$\Rightarrow$ $5y - 6x = 14.$

Example 28:

What relations must hold between the coefficients of the equations (i) $ax^2 + by^2 + cx + cy = 0$. (ii) $ay^2 + bxy + dy + cx = 0$, so that each of them may represent a pair of straight lines ?

Solution:

The given equation is

(i) $ax^2 + by^2 + cx + cy = 0$

If it represents 2 straight lines, D must be zero.

Hence $\Delta = a.b.0 + 2.\frac{c}{2}.\frac{c}{2}0 - a.\left(\frac{c}{2}\right)^2 - b\left(\frac{c}{2}\right)^2 - 0 = 0$

or $-\frac{ac^2}{4} - \frac{bc^2}{4} = 0$ or $c^2\ (a+b) = 0$.

So either c = 0 or (a + b) = 0 or c (a + b) = 0.

(ii) $D = 0.a.0 + 2.\frac{d}{2}.\frac{c}{2}.\frac{b}{2} - 0\left(\frac{d}{2}\right)^2 - a\left(\frac{c}{2}\right)^2 - 0\left(\frac{b}{2}\right)^2 = 0$

or $\frac{bdc}{4} - \frac{ac^2}{4} = 0$ or c (bd – ac) = 0.

Hence, either c = 0 or bd – ac = 0.

Example 29(a):

What loci are represented by the equation $x^2 - y^2 = 0$.

Solution:

The given equation is

$x^2 - y^2 = 0$ or $(x - y)(x + y) = 0$.

Hence the equation represents 2 straight lines

x – y = 0 and x + y = 0.

Example 29(b):

What loci are represented by the equation $x^2 - xy = 0$.

Solution:

The given equation is

$x^2 - xy = 0$ or $x(x - y) = 0$.

So the given equation represents two straight lines as x = 0, i.e., y-axis and (x – y) = 0.

Example 29(c):

What loci are represented by the equation $x^3 + y^3 = 0$.

Solution:

Hence, the equation represents three straight lines first being $x + y = 0$, i.e., passing through origin and inclined at an angle of 135° to x-axis, the other two being imaginary as $x^2 - xy + y^2 = 0$ has no real roots, but it represents two straight lines, passing through origin, being homogeneous in second degree.

Example 29(d):

What loci are represented by the equation $x^2 + y^2 = 0$.

Solution:

The given equation is

$$x^2 + y^2 = 0.$$

As the expression on L.H.S. cannot be factorized involving real factors and is a homogeneous equation in second degree, so represents 2 imaginary lines passing through origin.

In real variables, the relation is only possible when both $x = 0$ and $y = 0$ are satisfied; so it represents a point, i.e., origin.

Example 29(e):

What loci are represented by the equation $x^2y = 0$.

Solution:

So the equation represents three straight lines, as $x = 0$ and $y = 0$, the first two coinciding representing axis of y and the third one representing axis.

Example 30(a):

What loci are represented by the equation $(x^2 - 1)(y^2 - 4) = 0$.

Solution:

$$(x^2 - 1)(y^2 - 4) = 0$$

$$\Rightarrow \quad (x + 1)(x - 1)(y + 2)(y - 2) = 0$$

Hence, the given equation represents four st. lines i.e. $x = \pm 1$ (Par to axis) and $y = \pm 2$ (Parallel to x-axis)

Example 30(b):

Find the equation to the pair of straight lines joining the origin to the intersections of the straight line $y = mx + c$ and the curve

$x^2 + y^2 = a^2.$

Prove that they are at right angles, if $2c^2 = a^2(1 + m^2)$.

Solution:

Straight line is given as $y = mx + c$ **or** $\frac{y}{c} - \frac{mx}{c} = 1.$ **...(1)**

The curve is given as $x^2 + y^2 = a^2$ **or** $x^2 + y^2 - a^2 = 0.$ **...(2)**

Making (2) homogeneous with the help of (1), we get

$$\Rightarrow \quad x^2 + y^2 - a^2\left(\frac{y}{c} - \frac{mx}{c}\right)^2 = 0$$

$$x^2 + y^2 - a^2\left[\frac{y^2}{c^2} + \frac{m^2x^2}{c^2} - \frac{2myx}{c^2}\right] = 0$$

$$\Rightarrow \quad c^2x^2 + c^2y^2 - a^2y^2 - a^2m^2x^2 + 2a^2myx = 0.$$

$$\Rightarrow \quad x^2(c^2 - a^2m^2) + y^2(c^2 - a^2) + 2a^2myx = 0. \quad ...(3)$$

This equation represents the two lines joining the meeting point of (1) and (2) with the origin.

If the lines represented by (3) are at right angles, then

$$(c^2 - a^2m^2) + (c^2 - a^2) = 0$$

[as the sum of the coefficients of x and y^2 **must be zero]**

$$2c^2 = a^2(1 + m^2).$$

Example 31:

Prove that the straight lines joining the origin to the points of intersection of the straight line $kx + hy = 2hk$ *with the curve*

$(x - h)^2 (y - k)^2 = c^2$ *are at right angles if,* $h^2 + k^2 = c^2$.

Solution:

Then given is $kx + hy = 2hk$ **or** $\frac{x}{2h} + \frac{y}{2k} = 1.$ **...(1)**

The given curve is $(x - h)^2 + (y - k)^2 = c^2$

$\Rightarrow x^2 + y^2 - 2hx - 2ky + h^2 + k^2 - c^2 = 0.$ **...(2)**

Making (2) homogeneous with the help of (1), we get

$$x^2 + y^2 - 2hx\left(\frac{x}{2h} + \frac{y}{2k}\right) - 2ky\left(\frac{x}{2h} + \frac{y}{2k}\right) + (h^2 + k^2 - c^2)\left(\frac{x}{2h} + \frac{y}{2k}\right)^2 = 0. \quad ...(3)$$

Coefficient of x^2 in (3) is $= \dfrac{h^2+k^2-c^2}{4h^2}$

Coefficient of y^2 in (3) is $= \dfrac{h^2+k^2-c^2}{4k^2}$.

If the lines represented by (3) are perpendicular to each other, then

$$\frac{h^2+k^2-c^2}{4h^2}+\frac{h^2+k^2-c^2}{4k^2}=0$$

$$\Rightarrow \quad \left(h^2+k^2-c^2\right)\left(\frac{1}{4h^2}+\frac{1}{4k^2}\right)=0.$$

As $\left(\dfrac{1}{4h^2}+\dfrac{1}{4k^2}\right)$ cannot be zero, being sum of two squares, hence.

Example 32 :

Show that the equation $bx^2 - 2xy + ay^2 = 0$ represents a pair of straight lines which are at right angles to the pair given by the equation $ax^2 + 2hxy + by^2 = 0$.

Solution :

First equation is given as, $bx^2 - 2hxy + ay^2 = 0$ ...(1)

Let the equation represented by (1) be

$y = m_1x$ ...(2)

and $y = m_2x$. ...(3)

Then $(y - m_1x)(y - m_2x) = 0$ must be same as (1)

or $y^2 - (m_1 + m_2)xy + m_1m_2x^2 = 0$ must be identical with $bx^2 - 2hxy + ay^2 = 0$.

Comparing the coefficients, $\dfrac{1}{a}=\dfrac{-(m_1+m_2)}{-2h}=\dfrac{m_1 m_2}{b}$.

Hence $m_1 m_2 = \dfrac{b}{a}$ ...(4)

$m_1 + m_2 = \dfrac{2h}{a}$. ...(5)

The line perpendicular to (2) is $y=-\dfrac{1}{m^2}x$ or $m_2y + x = 0$.

Combined equation of these two is $(m_1y + x)(m_2y + x) = 0$

$\Rightarrow \quad m_1m_2y_2 + (m_1 + m_2)xy + x^2 = 0.$

Substituting the values from (4) and (5), we get

$$\frac{b}{a}y^2+\frac{2h}{a}xy+x^2=0$$

$$\Rightarrow \quad by^2 + 2hxy + ax^2 = 0$$

which is same as the given second equation.

Example 33 :

If pairs of straight lines $x^2 - 2pxy - y^2 = 0$ and $x^2 - 2qxy - y^2 = 0$ be such that each pair bisects the angles between the other pair, prove the $pq = 1$.

Solution :

The given equation is $x^2 - 2pxy - y^2 = 0$. ...(1)

As the equation of the pairs of the lines bisecting the angles between the lines $ax^2 + 2hxy + by^2 = 0$ is $\frac{x^2-y^2}{a-b}=\frac{xy}{h}$.

Hence the equation of the lines bisecting the angles between (1) will be

$$\frac{x^2-y^2}{1-(-1)}=\frac{xy}{-p}.$$

or $\quad px^2 + 2xy - py^2 = 0.$ ...(2)

The given equation of the bisector is $x^2 - 2qxy - y^2 = 0.$ ...(3)

As (2) and (3) represent the same straight lines, comparing the coefficients,

$$\frac{p}{1}=\frac{2}{-2q} \text{ or } pq = -1.$$

Example 34(a):

Prove that the pair of lines $a^2x^2 + 2h(a + b)xy + b^2y^2 = 0$ is equally inclined to the pair $ax^2 + 2hxy + by^2 = 0$.

Solution:

If each line of the pair of straight lines be equally inclined to the lines of the other pair of straight lines, they will have a common bisector.

The first equation is given as

$$a^2x^2 + 2h(a + b)xy + b^2y^2 = 0. \quad ...(1)$$

The equation of the bisector will be

$$\frac{x^2-y^2}{a^2-b^2}=\frac{xy}{h(a+b)} \text{ or } \frac{x^2-y^2}{a-b}=\frac{xy}{h}. \quad ...(2)$$

The other equation is given as $ax^2 + 2hxy + by^2 = 0$. ...(3)

The equation of bisector clearly $\dfrac{x^2 - y^2}{a - b} = \dfrac{xy}{h}$ which is same as (2), i.e., bisector of (1).

Example 34(b):

Prove that the equation to the bisectors of the angle between the straight lines $ax^2 + 2hxy + by^2 = 0$ is $h(x^2 - y^2) + (b - a)xy = (ax^2 - by^2)\cos\omega$, the axes being inclined at an angle ω.

Solution:

Let X'OX and YOY' be the axes inclined at an angle ω.CD and AB be two straight lines represented by given equation $ax^2 + 2hxy + by^2 = 0$ inclined at angles θ_1 and θ_2 to x-axis respectively. Let PQ and RS be the bisectors of the angles between the lines AB and CD. If PQ is inclined at an angle θ to x-axis, RS will be inclined at $(\pi/2 + \theta)$.

Again $\theta + \theta_1 + \dfrac{1}{2}(\theta_1 - \theta_2) = \dfrac{\theta_1 + \theta_2}{2}$

or $\qquad 2\theta = (\theta_1 + \theta_2)$ ...(1)

Let CD be represented by $y = m_1x$ and AB by $y = m_2x$. Hence the combined equation will be $(y - m_1x)(y - m_2x) = 0$.

Comparing with given equation, we get

$$m_1 + m_2 = -\frac{2h}{b} \text{ and } m_1 m_2 = \frac{a}{b}.$$

Now $\tan 2\theta = \tan(\theta_1 + \theta_2) = \dfrac{\tan\theta_1 + \tan\theta_2}{1 - \tan\theta_1 \tan\theta_2}$

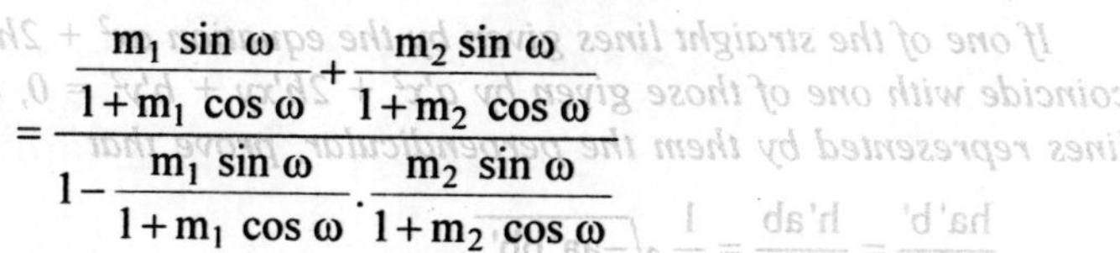

$$= \frac{\dfrac{m_1 \sin\omega}{1 + m_1\cos\omega} + \dfrac{m_2\sin\omega}{1 + m_2\cos\omega}}{1 - \dfrac{m_1\sin\omega}{1 + m_1\cos\omega}\cdot\dfrac{m_2\sin\omega}{1 + m_2\cos\omega}}$$

$$\therefore \left[\begin{array}{c}\text{if the line} - m_1x \text{ makes an angle } \theta_1 \text{ from } x - \text{axis} \\ \tan q_1 = \dfrac{m_1\sin\omega}{1 + m_1\cos\omega}\end{array}\right]$$

$$= \frac{2m_1 m_2 \sin\omega \cos\omega + (m_1 + m_2) \sin\omega}{1 + (m_1 + m_2)\cos\omega - m_1 m_2 (\sin^2\omega - \cos^2\omega)}$$

$$\therefore \quad \tan 2\theta = \frac{2a \sin\omega \cos\omega - 2h \sin\omega}{b - 2h\cos\omega - a(\sin^2\omega - \cos^2\omega)}$$

$$\left(\because \quad m_1 m_2 = -\frac{2h}{b}, m_1 m_2 = \frac{a}{b}\right) \qquad ...(2)$$

Again let PQ be y = mx. ...(3)

As PQ is inclined at an angle of θ to x-axis $\tan\theta = \dfrac{m \sin\omega}{1 + m\cos\omega}$

$$\therefore \tan 2\theta = \frac{2\tan\theta}{1 - \tan^2\theta} = \frac{2m\sin\omega/(1 + m\cos\omega)}{1 - \{m\sin\omega/(1 + m\cos\omega)\}^2}.$$

$$= \frac{2\frac{y}{x}\sin\omega \Big/ \left(1 + \frac{y}{x}.\cos\omega\right)}{1 - \left\{\frac{y}{x}\sin\omega \Big/ \left(1 + \frac{y}{x}\cos\omega\right)\right\}} \quad \left(\because \quad m = \frac{y}{x} \text{ by (3)}\right)$$

$$= \frac{2y\sin\omega(x + y\cos\omega)}{(x + y\cos\omega)^2 - (y\sin\omega)^2} \qquad ...(4)$$

Equating R.H.S. of (2) and (4), as L.H.S. are same, we get

$$\frac{2a\sin\omega\cos\omega - 2h\sin\omega}{b - 2h\cos\omega - a(\sin^2\omega - \cos^2\omega)} = \frac{2y\sin\omega(x + y\cos\omega)}{(x + y\cos\omega)^2 - (y\sin\omega)^2}$$

Simplifying we get

$$h(x^2 - y^2) - (b - a)xy = (ax^2 - by^2)\cos\omega.$$

Example 35:

If one of the straight lines given by the equation $ax^2 + 2hxy + by^2 = 0$, coincide with one of those given by $a'x^2 + 2h'xy + b'y^2 = 0$, and the other lines represented by them the perpendicular, prove that

$$\frac{ha'b'}{b' - a'} = \frac{h'ab}{b - a} = \frac{1}{2}\sqrt{-aa'bb'}.$$

Solution:

The equations are given as

$$ax^2 + 2hxy + by^2 = 0 \qquad ...(1)$$

and $\quad a'x^2 + 2h'xy + b'y^2 = 0.$ ...(2)

Let the equations represented by (2) be

$$y = m_1x \quad ...(3)$$

and $\quad y = m_2x.$ ...(4)

Then combined equation of (3) and (4) will be given as

$$(y - m_1x)(y - m_2x) = 0$$

or $\quad y^2 - (m_1 + m_2)xy + m_1m_2x^2 = 0.$ ...(5)

As (a) and (5) represent the same equation hence comparing the coefficients $\frac{1}{b} = -\frac{m_1 + m_2}{2h} = \frac{m_1 m_2}{a}$.

$$\therefore \quad m_1 + m_2 = -\frac{2h}{b} \quad ...(6)$$

and $\quad m_1 m_2 = \frac{a}{b}.$ ...(7)

As (2) represents a straight line coinciding with (3) and other perpendicular to (4), we must have (2) similar to $(y - m_1x)(m_2y + x) = 0$.

Comparing with (2), we get

$$\frac{b'}{m_2} = \frac{2h'}{1 - m_1 m_2} = \frac{a'}{-m_1} \text{ or } \frac{m_1}{m_2} = -\frac{a'}{b'} \quad ...(8)$$

and

$$\frac{m_1 m_2 - 1}{m_2} = -\frac{2h'}{b'}$$

$$\Rightarrow \quad m_1 - \frac{1}{m_2} = -\frac{2h'}{b'} \quad ...(9)$$

Solving (6) and (8), we get $m_1 = \frac{-2a'h}{b(a'-b')}$ and $m_2 = \frac{2b'h}{b(a'-b')}$

Substituting in (7), we get $\frac{-2a'h}{b(a'-b')} \cdot \frac{2b'h}{b(a'-b')} = \frac{a}{b}$

$$\Rightarrow \quad \frac{4h^2a'b'}{(a'-b')^2} = -ab$$

$$\Rightarrow \quad \frac{4h^2a'^2b'^2}{(b'-a')^2} = -aba'b'$$

$$\Rightarrow \quad \frac{ha'b'}{b'-a'} = \frac{1}{2}\sqrt{(-aba'b')}.$$

Similarly by (7) and (8), $m_1^2 = -\frac{aa'}{bb'}$ and $\frac{1}{m_2^2} = \left(-\frac{a'b}{ab'}\right)$. ...(10)

Putting in (9), $\sqrt{\left(-\frac{aa'}{bb'}\right)} - \sqrt{-\left(\frac{a'b}{ab'}\right)} = -\frac{2h'}{b'}$

$$\Rightarrow \quad \sqrt{\left(-\frac{a'}{b'}\right)}\left[\frac{\sqrt{a}}{\sqrt{b}} - \frac{\sqrt{b}}{\sqrt{a}}\right] = -\frac{2h'}{b'}$$

$$\Rightarrow \quad \sqrt{\left(-\frac{a'}{b'}\right)}\left[\frac{a-b}{\sqrt{(ab)}}\right] = -\frac{2h'}{b'}$$

$$\Rightarrow \quad \sqrt{(-a'abb)}.(a-b) = -2h'ab$$

$$\Rightarrow \quad \frac{h'ab}{b-a} = \frac{1}{2}\sqrt{(-aa'bb')} \qquad ...(11)$$

By (10) and (11), we get the required result.

Example 36(a):

Show that the distance between the points of intersection of the straight line x cos α + y sin α − p = 0, with the straight lines $ax^2 + 2hxy + by^2$ = 0 is $\frac{2p\sqrt{h^2 - ab}}{b\cos^2\alpha - 2h\cos\alpha\sin\alpha + \alpha\sin^2\alpha}$

Deduce the area of the triangle formed by them.

Solution:

The equations are given as

$$x\cos\alpha + y\sin\alpha - p = 0 \qquad ...(1)$$

and $$ax^2 + 2hxy + by^2 = 0 \qquad ...(2)$$

To above them, putting the value of x from (1) in (2), we get

$$a.\left(\frac{p - y\sin\alpha}{\cos\alpha}\right) + 2h.\frac{p - y\sin\alpha}{\cos\alpha}y + by^2 = 0$$

$$\Rightarrow \quad a(p - ya)^2 + 2h\cos\alpha(p - y\sin\alpha).y + by^2\cos^2\alpha = 0$$

$$\Rightarrow \quad y^2[a\sin^2\alpha - 2h\cos\alpha\sin\alpha + b\cos^2\alpha]$$

$$- 2y(ap\sin\alpha - ph\cos\alpha) + ap^2 = 0$$

This being a quadratic equation will give us 2 values of y.

Let these values by y_1 and y_2, then

$$y_1 + y_2 = \frac{2p\,(a \sin\alpha - h\cos\alpha)}{a\sin^2\alpha - 2h\cos\alpha\sin\alpha + b\cos^2\alpha} \qquad ...(3)$$

and $$y_1 y_2 = \frac{ap^2}{a\sin^2\alpha - 2h\cos\alpha\sin\alpha + b\cos^2\alpha} \qquad ...(4)$$

If x_1 and x_2 be the corresponding values of abscissa for y_1 and y_2, then the points $(x_1, y_1$ and (x_2, y_2) will be the point of intersection of (1) and (2). So these will satisfy (1).

$$\therefore \quad x_1 \cos\alpha + y_1 \sin\alpha - p = 0$$

and $$x_2 \cos\alpha + y_2 \sin\alpha - p = 0.$$

Substracting, we get, $\cos\alpha\,(x_1 - x_2) + \sin\alpha\,(y_1 - y_2) = 0.$ ...(5)

If d be the distance between the points (x_1, y_1) and (x_2, y_2), then $d^2 =$ $\sqrt{[(x_1 - x_2)^2 + (y_1 - y_2)^2]} = \sqrt{[(y_1 - y_2)^2 \ \tan^2\alpha + (y_1 - y_2)^2]}$

$$\left\{\therefore \ (x_1 - x_2) = -\frac{(y_1 - y_2)}{\cos\alpha}\right\} \text{ by (5)}$$

$$= \sqrt{[(y_1 - y_2)^2 \ (\tan^2\alpha + 1)]} = \sec\alpha\sqrt{[(y_1 + y_2)^2\, 4\,(y_1\, y_2)]}$$

$$= \sec\alpha\sqrt{\left[\left\{\frac{2p\ (a\sin\alpha - h\cos\alpha)}{a\sin^2 a - 2h\cos a\sin a + b\cos^2 a}\right.\right. - \left.\frac{4ap^2}{a\sin^2\alpha - 2h\cos\alpha\sin\alpha + b\cos^2\alpha}\right]}$$

$$= \frac{\sec\alpha\ 2p}{a\sin^2\alpha - 2h\sin\alpha\cos\alpha + b\cos^2\alpha} \times [a^2\sin^2\alpha + h^2\cos^2\alpha - 2ah\sin\alpha\cos\alpha)] - [a^2\sin^2\alpha - 2ha\sin\alpha\cos\alpha + ab\cos^2\alpha)].$$

$$= \frac{2p\sec\alpha\sqrt{\{\cos^2\alpha\,(h^2 - ab)\}}}{a\sin^2\alpha - 2h\sin\alpha\cos\alpha + b^2\cos\alpha}$$

$$= \frac{2p\sec\alpha\cos\alpha\sqrt{\{\cos^2\alpha\,(h^2 - ab)\}}}{a\sin^2\alpha - 2h\sin\alpha\cos\alpha + a\sin^2\alpha}$$

$$= \frac{2p\sqrt{\{(h^2 - ab)\}}}{b\cos^2\alpha - 2h\sin\alpha\cos\alpha + a\sin^2\alpha}$$

If the points of intersection be A and B, the line BA will be x cos α + y sin α – p = 0. The length of perpendicular on AB from the third vertex of the triangle which is origin, i.e., (0, 0) will be p (as the equation is in standard form in which p is the length of perpendicular from origin on the line).

$$\text{So its area} = \frac{1}{2}\cdot\frac{p^2\sqrt{\{(h^2-ab)\}}}{b\cos^2\alpha - 2h\cos\alpha\sin\alpha + a\sin^2\alpha}$$

$$= \frac{p^2\sqrt{\{(h^2-ab)\}}}{b\cos^2\alpha - 2h\cos\alpha\sin\alpha + a\sin^2\alpha}. \qquad \textbf{Ans.}$$

Example 36(b):

Prove that the general equation $ax^2 + 2hxy + by^2 + 2gx + 2fy + c = 0$, represents two parallel straight lines if $h^2 = ab$ and $bg^2 = af^2$ Prove also that the distance between them is $2\sqrt{\dfrac{g^2-ac}{a(a+b)}}$.

Solution:

The given equation is

$$ax^2 + 2hxy + by^2 + 2gx + 2fy + c = 0. \qquad ...(1)$$

Let the two equations represented by (1) be respectively

$$\left(\sqrt{a}.x + \sqrt{b}.y + l\right) = 0. \qquad ...(2)$$

and $$\left(\sqrt{a}.x + \sqrt{b}.y + m\right) = 0.$$

Which is identical to (1) and as the coefficients of x^2 and y^2 are equal. Hence equating different coefficients

$$2h = 2\sqrt{(ab)} \qquad ...(4)$$

$$2g = \sqrt{a}\,(l+m) \qquad ...(5)$$

$$2f = \sqrt{b}\,(l+m) \qquad ...(6)$$

and $$c = l\,m \qquad ...(7)$$

By (4), we get $h^2 = ab$

Dividing (5) by (6), we get $\dfrac{g}{f} = \dfrac{\sqrt{a}}{\sqrt{b}}$ or $\dfrac{g^2}{f^2} = \dfrac{a}{b}$.

Hence $bg^2 = af^2$.

Again, if p and p' be the lengths of perpendiculars from origin on (2) and (3), then the distance between them is p – p'.

$$\text{So } p - p' = \frac{1}{\sqrt{(a+b)}} - \frac{m}{\sqrt{(a+b)}} = \frac{(l-m)}{\sqrt{(a+b)}} = \sqrt{\left\{\frac{(l+m)^2 - 4lm}{a+b}\right\}}.$$

Example 37:

Prove that the product of the perpendiculars let fall from the point (x', y') upon the pair of straight lines

$ax^2 + 2hxy + by^2 = 0$ is

$$\frac{ax^2 + 2hx'y' + by}{\sqrt{(a-b)^2 + 4h^2}}.$$

Solution:

Let the line $ax^2 + 2hxy + by^2 = 0$ represent the two straight lines

$$y - m_1x = 0 \qquad ...(1)$$

and

$$y - m_2x = 0. \qquad ...(2)$$

$$m_1 m_2 = \frac{a}{b} \qquad ...(4)$$

If p_1 and p_2 be the lengths of perpendiculars from (x_1, y_1) on (1) and (2) respectively, then we have

$$p_1 p_2 = \frac{y_1 - m_1 x_1}{\sqrt{(1+m_1^2)}} \cdot \frac{y_1 - m_2 x_1}{\sqrt{(1+m_2^2)}}$$

$$= \frac{y_1^2 - x_1 y_1 (m_1 + m_2) + m_1 m_2 x_1^2}{y + m_1^2 + m_2^2 + (m_1 m_2)^2}$$

$$= \frac{y_1^2 - x_1 y_1 (m_1 + m_2) + m_1 m_2 x_1^2}{1 + (m_1 + m_2)^2 - 2m_1 m_2 + (m_1 m_2)^2}$$

Putting the values of $m_1 + m_2$ and m_1m_2 from (3) and (4),

$$p_1p_2 = \frac{y_1^2 - \left(-\frac{2h}{b}\right) x_1 y_1 + \left(\frac{a}{b}\right) x_1^2}{\sqrt{\left(1 + -\frac{2h}{b}^2\right) - 2\frac{a}{b} + \left(\frac{a}{b}\right)^2}}$$

$$= \frac{by_1^2 + 2hx_1y_1 + ax_1^2}{\sqrt{\{b^2 + 4h^2 - ab + a^2\}}} = \frac{ax_1^2 + 2hx_1 y_1 + by_1^2}{\sqrt{\{(a-b)^2 + 4h^2\}}}.$$

Example 38(a):

What loci are represented by the equation $(x-a)^2 - y^2 = 0$.

Solution:

The given equation is

$$(x - a)^2 - y^2 = 0.$$

Factorizing, we get $(x - a - y)(x - a + y) = 0$.

Hence the given equation represents two straight lines as $x - y - a = 0$ and $x + y - a = 0$. The first cutting an intercept of a and –a from the axes and the second of a and a.

Example 38(b):

What loci are represented by the equation

$(x + y)^2 - c^2 = 0$ $(x + y - c)(x + y + c) = 0$.

Solution:

The given equation is

$$(x + y)^2 - c^2 = 0$$

or $$(x + y - c)(x + y + c) = 0.$$

Hence the equation represents 2 straight $x + y - c = 0$ and $x + y + c = 0$, first cutting the intercepts of c from both the axes second –c from both the axes.

Example 39(a):

What loci are represented by the equation $r = a \sec(\theta - \alpha)$.

Solution:

The given equation is

$$r = a \sec(\theta - \alpha) \text{ or } r = \frac{a}{\cos(\theta-\alpha)} \text{ or } r\cos(\theta - \alpha) = a$$

$$\Rightarrow \quad r(\cos\theta\cos\alpha + \sin\theta\sin\alpha) = a$$

$$\Rightarrow \quad r\cos\theta\cos\alpha + r\sin\theta\sin\alpha = a$$

$$\Rightarrow \quad x\cos\alpha + y\sin\alpha = a \qquad (\therefore r\cos\theta = x,\ r\sin\theta = y).$$

This is a standard form of a straight line. Hence the given equation represents a straight line.

Example 39(b):

What loci are represented by the equation

$(x^2 - 1)^2 + (y^2 - 4)^2 = 0$.

Solution:

The given equation is

$$(x^2 - 1)^2 + (y^2 - 4)^2 = 0.$$

As the sum of two squares cannot be equal to zero unless each of them is equal to zero, hence we must have both

$$x^2 - 1 = 0$$

and $\quad y^4 - 4 = 0$

$\Rightarrow \quad (x - 1)(x + 1) = 0$

and $\quad (y - 2)(y + 2) = 0.$

Hence, we get 4 different combinations, each representing a point. These points are (1, 2), (1, –2), (–1, 2) and (–1, 2).

Example 40:

What loci are represented by the equation

$$(y - mx - c)^2 + (y - mx - c')^2 = 0.$$

Solution:

The given equation is

$(y - mx + c)^2 + (y - mx - c')^2 = 0$

$\quad (y - mx - c)^2 = 0$

or $\quad y - mx - c = 0$ ...(1)

and $\quad (y - m'x - c)^2 = 0$

or $\quad y - m'x - c' = 0.$ ...(2)

As both the equations (1) and (2) must be satisfied, the equation represents a point which is the point of intersection of (1) and (2). Solving (1) and (2), we get the point as

$$\left(\frac{c'-c}{m-m,}, \frac{mc'-m'c}{m-m'}\right).$$

Example 41:

What loci are represented by the equation

$$(x^2 - a^2)^2 (x^2 - b^2)^2 + c^4 (y^2 - a^2)^2 = 0.$$

Solution:

The given equation is

$$(x^2 - a^2)^2 (x^2 - b^2)^2 + c^4 (y^2 - a^2)^2 = 0.$$

This being the sum of two perfect squares, each term must be zero. Hence, we get

$$(x^2 - a^2)(x^2 - b^2)^2 (x^2 - b^2)^2 + c^4 (y^2 - a^2) = 0.$$

$\Rightarrow$ $(x - a)(x + a)(x - b)(x + b) = 0$ ...(1)

and $c^4 (y^2 - a^2)^2 = 0$

$\Rightarrow$ $c^2 (y^2 - a^2) = 0$

or $c^2 (y + a)(y - a) = 0.$ ...(2)

Equation no. (1) holds good for $x = \pm a$ or $x = \pm b$.

Equation no. (2) is satisfied by $y = \pm a$.

As both of these should be simultaneously satisfied, the given line represents 8 points which we get as a result of different combinations of (1) and (2), namely $(\pm a, \pm a)$, $(\pm b, \pm a)$.

Example 42(a):

Show also that the pair $ax^2 + 2hxy + by^2 + \lambda (x^2 + y^2) = 0$ *is equally inclined to the same pair.*

Solution:

The given equation is

$$ax^2 + 2hxy + by^2 + \lambda (x^2 + y^2) = 0 \quad ...(1)$$

$$(a + \lambda) x^2 + 2hxy + (b + \lambda) y^2 = 0.$$

The equation of bisector will be

$$\frac{x^2 - y^2}{(a+\lambda)-(b+\lambda)} = \frac{xy}{h} \text{ or } \frac{x^2 - y^2}{a-b} = \frac{xy}{h}$$

This equation is independent of λ. Hence (2) has the same bisector for all values of λ.

Example 42(b):

Show that the equation $\cos 3\alpha (x^3 - 3xy^2) + \sin 3\alpha (y^3 - 3x^2y) + 3a (x^2 + y^2) - 4a^3 = 0$ *represents three straight lines forming an equilateral triangle. Prove also that its area is* $3\sqrt{3a^2}$.

Solution:

The given equation is

$$\cos 3\alpha (x^3 - 3xy^2) + \sin 3\alpha (y^3 - 3x^2) + 3a (x^2 + y^2) - 4a^3 = 0$$

Changing in polar co-ordinates by putting $x = r\cos\theta$, $y = r\sin\theta$, we get

$$\cos 3\alpha (r^3 \cos^3\theta - 3r^3 \cos\theta \sin^2\theta) + \sin 3\alpha$$

$$(r^3 \sin^3\theta - 3 \cos^2 \theta \sin \theta) + 3a (r^2 \cos^2 \theta + r^2 \sin^2\theta) - 4a^3 = 0$$

$$\Rightarrow r^3 \cos 3\alpha \cos \theta (\cos^2 \theta - 3 \sin^2 \theta)$$

$$+ r^3 \sin 3\alpha \sin \theta (\sin^2 - 3 \cos^2\theta) + 3ar^2 - 4a^3 = 0$$

As $\cos^2 \theta - \sin^2 \theta = 4 \cos^2\theta - 3$

and $\sin^2\theta - 3 \cos^2 \theta = -(3 - 4 \sin^2 \theta)$,

we get $r^3 \cos 3\alpha.[4 \cos^3 \theta - 3 \cos \theta]$

$$-r^2 \sin 3\alpha [3 \sin \theta - 4 \sin^3 \theta] + 3ar^2 - 4a^3 = 0$$

$$\Rightarrow r^3 [\cos 3\alpha \cos 3\theta - \sin 3\alpha \sin 3\theta] + 3ar^2 - 4a^3 = 0$$

$$\Rightarrow r^3 \cos (3\alpha + 3\theta) + 3ar^2 - 4a^3 = 0$$

$$\Rightarrow \cos (3\alpha + 3\theta) = 4 \left(\frac{a}{r}\right)^3 - 3\left(\frac{a}{r}\right).$$

As R.H.S. is of the form $4 \cos^3 \lambda - 3 \cos l)$, the above equation may have 3 different solutions between 0 and 360° as

$$\frac{a}{r} = \cos (a + q) \qquad ...(1)$$

$$\frac{a}{r} = \cos\left(\alpha + \theta + \frac{2\pi}{3}\right) \qquad ...(2)$$

and
$$\frac{a}{r} = \cos\left(\alpha + \theta + \frac{4\pi}{3}\right). \qquad ...(3)$$

Hence, the given line represent 3 straight lines whose equations are given by (1), (2) and (3). The acute angle between (1, 2), (2, 3) and (3, 1) is clearly supplementary of 2p/3 or 60°.

Hence, if they meet at A, B and C respectively, all the angles D,A,B,C are 60° each. Therefore, ABC is an equilateral triangle.

Again (1) is $\frac{a}{r} = \cos(\alpha + \theta)$.

$$\Rightarrow a = r (\cos \alpha \cos \theta - \sin \alpha \sin \theta)$$

$$\Rightarrow a = r \cos \theta \cos \alpha - r \sin \theta \sin \alpha$$

$$\Rightarrow x \cos \alpha - y \sin \alpha = a \qquad ...(4)$$

[as $x = r \cos \theta$ and $y = r \sin \theta$].

Similarly, (2) and (3) on changing into Cartesian co-ordinates become

$$x \cos\left(\alpha + \frac{2\pi}{3}\right) - y \sin\left(\alpha + \frac{2\pi}{3}\right) = a \qquad ...(5)$$

and $$x\cos\left(\alpha+\frac{4\pi}{3}\right)-y\sin\left(\alpha+\frac{4\pi}{3}\right)=a \quad ...(6)$$

respectively.

To get the point of intersection of (2) and (3); i.e., B, simplifying (5), we get

$$x\cos\alpha\cos\frac{2\pi}{3}-x\sin\alpha\sin\frac{2\pi}{3}-y\sin\alpha\cos\frac{2\pi}{3}-y\cos\alpha\sin\frac{2\pi}{3}=a$$

$$\Rightarrow -x\cos\alpha.\frac{1}{2}-x\sin\alpha\frac{\sqrt{3}}{2}+y\sin\alpha\frac{1}{2}-y\cos\alpha\frac{\sqrt{3}}{2}=a$$

$$\Rightarrow -\cos\alpha - x\sqrt{3}\sin\alpha + y\sin\alpha - y\sqrt{3}\cos\alpha = 2a. \quad ..(7)$$

Similarly, simplifying (6), we get

$$-x\cos\alpha + x\sqrt{3}\sin\alpha + y\sin\alpha - y\sqrt{3}\cos\alpha = 2a. \quad ...(8)$$

Adding (7) and (8), we get

$$-2x\cos\alpha + 2y\sin\alpha = 4a \text{ or } -x\cos\alpha + y\sin\alpha = 2a. \quad ..(9)$$

Subtracting (8) from (7), we get

$$-2x\sqrt{3}\sin\alpha - 2y\sqrt{3}\cos\alpha = 0$$

$$\Rightarrow \quad x\sin a + y\cos a = 0. \quad ...(10)$$

Solving (9) and (10), we get $x = -2a\cos\alpha$, $y = 2a\sin\alpha$.

Hence the co-ordinates of B are $(-2a\cos\alpha, 2a\sin\alpha)$.

Equation of AC is given by (4).

Perpendicular distance of B from AC

$$=\frac{-2a\cos^2\alpha-2a\sin^2\alpha-a}{\sqrt{(\cos^2\alpha+\sin^2\alpha)}}=3a \text{ (omitting – ve sign).}$$

As the triangle ABC is equilateral one, Its area $=\dfrac{(3a)^2}{\sqrt{3}}=3\sqrt{3a}^{2}$.

[∴ if p is the height of an equation triangle, its area will be $p^2/\sqrt{3}$].

Example 43:

Prove that two of the lines represented by the equation $ax^4 + bx^3y + cx^2y^2 + dxy^3 + ay^4 = 0$ will bisect the angles between the other two if $c = 6a = 0$ and $b + d = 0$.

Solution:

The equation is given as

$$ax^4 + bx^3y + cx^2y^2 + dxy^3 + axy^3 + ay^4 = 0. \qquad ...(1)$$

The equation being homogeneous equation of fourth degree will represent 4 straight lines. Let one pair of these be

$$(px^2 + 2qxy + ry^2) \qquad ...(2)$$

The equations of the bisectors of the angles between the lines given by (2) will be $\dfrac{x^2 - y^2}{p - r} = \dfrac{xy}{q}$.

$$\theta\,(x^2 - y^2) - (p - r)\,xy = 0.$$

As (1) represents by hypothesis 2 pairs of straight lines, one pair of which is the bisector of the angles between the other set, we must have

$$ax^4 + bx^3y + cx^2y^2 + dxy^3 + ay^4 \equiv (px^2 + 2qxy + ry^2)\,[qx^2 - qy^2 - (p - r)\,xy].$$

Comparing the coefficient of different terms in order, we get

$$\frac{a}{pq} = \frac{b}{-p^2 + pr + 2q^2} = \frac{c}{-3q\,(r - p)} = \frac{d}{-2p^2 - pr + r^2} = -\frac{a}{qr}.$$

By 1 and 5 ratio, $r = -p$.

Putting in 2 and 4 ratios.

$$-\frac{b}{p^2 - p^2 = 2q^2} = \frac{d}{-2q^2 + p^2 + p^2}$$

or $\qquad b = -d$ or $b + d = 0$.

Again by 1 and 3 ratios

$$\frac{a}{pq} = \frac{c}{3p.(-p - p)} \text{ or } a = -\frac{c}{6} \text{ or } 6a + c = 0.$$

Example 44(a):

If the equation $ax^2 + 2hxy + by^2 + 2gx + 2fy + c = 0$, *represent a pair of straight lines, prove that the equation to the third pair of straight lines passing through the points where these meet the axes is*

$$ax^2 - 2hxy + by^2 + 2gx + 2fy + c + \frac{4fg}{c}xy = 0$$

Solution:

We are given the equations as

$$ax^2 + 2hxy + by^2 + 2gx + 2fy + c = 0. \quad ...(1)$$

The equation of y-axis is x = 0 and x-axis is y = 0. Hence the combined equation will be xy = 0. ...(2)

Equation of the curve passing through the point of intersection of (1) and (2) will be

$$ax^2 + 2hxy + by^2 + 2gx + 2fy + 2fy + 2\lambda xy = 0$$

$$\Rightarrow \quad ax^2 + 2(h + \lambda)\, xy + by^2 + 2gx + 2fy + c\ 0. \quad ...(3)$$

If (3) represent two straight lines, then its discriminant must be zero.

$$\text{So} \quad abc + 2.f.g.\ (h + \lambda) - af^2 - bg^2 - c\ (h + \lambda)^2 = 0$$

$$\Rightarrow \quad abc + 2fgh - af^2 - bg^2 - ch^2 + 2\lambda\ (fg - ch) - c\lambda^2 = 0 ...(4)$$

As (1) represents 2 straight lines

$$abc + 2fgh - af^2 - bg^2 - ch^2 = 0.$$

Putting in (4), we get $2\lambda\ (fg - ch) - cl = 0$ or $\lambda = \dfrac{2\,(fg-ch)}{c}$.

Putting in (3), we get

$$ax^2 + 2\left[h + \frac{2\,(fg-ch)}{c}\right]xy + by^2 + 2gx + 2fy + c = 0$$

$$\Rightarrow \quad ax^2 + \frac{4fg - 2ch}{c}xy + by^2 + 2gx + 2fy + c = 0$$

$$ax^2 - 2\ hxy + by^2 + 2gx + 2fy + c + \frac{4fg}{c}xy = 0.$$

Example 44(b):

Prove that the straight lines $ax^2 + 2hxy + by^2 = 0$, make equal angles with the axis of x if $h = a \cos \omega$, the axes being inclined at an angle ω.

Solution:

Let the equations represented by $ax^2 + 2hxy + by^2 = 0$ by

$$y - m_1x = 0$$

and $$y - m_2x = 0,$$

$$m_1 + m_2 = -2h/b$$

and $$m_1m_1 = a/b.$$

If the lines make equal angles with x-axis, one angle being θ, other must be (180 – θ).

So $$\tan\theta = \frac{m_1 \sin\omega}{1 + m_1 \cos\omega}, \qquad ...(1)$$

and $$\tan(180 - \theta) = \frac{m_2 \sin\omega}{1 + m_2 \cos\omega}. \qquad ...(2)$$

By (1) and (2), we get $$\frac{m_1 \sin\omega}{1 + m_1 \cos\omega} = -\frac{m_2 \sin\omega}{1 + m_2 \cos\omega}.$$

[as tan (180 – θ) = – tan θ]

$$\Rightarrow \quad m_1 + m_2 + 2m_1m_2 \cos\omega = 0. \qquad ...(3)$$

Putting the values of $m_1 + m_2$ and m_1m_2, we get

$$\frac{-2h}{b} + \frac{2a \cos w}{b} = 0$$

or $$h = a \cos\omega.$$

Example 45:

If the axes be inclined at an angle ω, show that the equation $x^2 + 2xy \cos\omega + y^2 \cos 2\omega = 0$, represents a pair of perpendicular straight lines.

Solution:

Let the given equation be

$$x^2 + 2xy \cos\omega + y^2 \cos 2\omega = 0 \qquad ...(1)$$

represent 2 straight lines $y = m_1x$ and $y = m_2x$. Then the combined equation will be

$$(y - m_1x)(y - m_2x) = 0. \qquad ...(2)$$

Comparing the coefficients of different terms in (1) and (2)

$$\frac{1}{m_1 m_2} = \frac{2\cos\omega}{-(m_1 + m_2)} = \frac{\cos 2\omega}{1}.$$

$$\therefore \quad m_1m_2 = \frac{1}{\cos 2\omega} \text{ and } m_1 + m_2 = \frac{2\cos\omega}{\cos 2\omega}$$

If lines given by (2) are perpendicular, then

$$1 + (m_1 + m_2)\cos\omega + m_1m_2 = 0.$$

Putting the values of $m_1 + m_2$ and m_1m_2 in L.H.S. we have

$$1 - \frac{2\cos\omega}{\cos 2\omega}\cos\omega + \frac{1}{\cos 2\omega}$$

$$= \frac{1}{\cos 2\omega}\left(\cos 2\omega - 2\cos^2\omega + 1\right) = 0 \quad (\because \cos 2\omega = 2\cos^2\omega - 1).$$

Hence, the lines represented by (1) are perpendicular to each other.

Example 46(a):

If the equation $ax^2 + 2hxy + by^2 + 2gx + 2fy + c = 0$, represent two straight lines, prove that the square of the distance of their point of intersection from the origin is $\frac{c(a+b) - f^2 - g^2}{ab - h^2}$.

Solution:

The given equation is

$$ax^2 + 2hxy + by^2 + 2gx + 2fy + c = 0 \quad ...(1)$$

be $\quad y - m_1x - c_1 = 0 \quad ...(2)$

and $\quad y - m_2x - c_2 = 0 \quad ...(3)$

The combined must be similar to (1), so

$ax^2 + 2hxy + by^2 + 2gx + 2fy + c \Rightarrow b(y - m_1x - c_1)(y - m_2x - c_2)$

(we multiply R.H.S. by b so as to make the coefficient of term say y^2 on R.H.S. equal to coefficient of y^2 on L.H.S. So the coefficients must also be equal). So equating other coefficients, we get

$$m_1 + m_2 = -\frac{2h}{b} \quad ...(4)$$

$$m_1 + m_2 = \frac{a}{b} \quad ...(5)$$

$$c_1 m_2 + c_2 m_1 = \frac{2g}{b} \quad ...(6)$$

$$c_1 + c_2 = -\frac{2f}{b} \quad ...(7)$$

and $$c_1c_2 = \frac{c}{b}. \quad ...(8)$$

The point of intersection of (2) and (3) will be given by

$$\frac{y}{m_1c_2 - m_2c_1} = \frac{x}{-c_1 + c_2} = \frac{1}{-m_2 + m_1}$$

Hence the co-ordinates of the point of intersection are

$$\left(\frac{c_2 - c_1}{m_1 - m_2}, \frac{m_1c_2 - m_2c_1}{m_1 - m_2}\right).$$

The square of distance of this point from the origin (0, 0) is

$$\left[\frac{c_2 - c_1}{m_1 - m_2}\right]^2 + \left[\frac{m_1 c_2 - m_2 c_1}{m_1 - m_2}\right]^2$$

$$= \frac{\left[(c_1 + c_2)^2 - 4c_1 c_2\right] + (m_1 c_2 + m_2 c_1)^2 - 4m_1 m_2 c_1 c_2}{(m_1 + m_2)^2 - 4m_1 m_2}$$

$$= \frac{\left[\frac{4f^2}{b^2} - \frac{4c}{b}\right] + \left[\frac{4g^2}{b^2} - \frac{4ac}{b^2}\right]}{\frac{4h^2}{b^2} - \frac{4a}{b}}$$

$$= \frac{f^2 - bc + g^2 - ac}{h^2 - ab} = \frac{f^2 + g^2 - c(b + a)}{h^2 - ab} = \frac{c(a + b) - f^2 - g^2}{ab - h^2}.$$

Example 46(b):

Show that two of the straight lines represented by the equation $ay^4 + bxy^3 + cx^2y^2 + dx^3y + ex^4 = 0$ will be at right angles if

$$(b + d)(ad + be) + (e - a)^2 (a + c + e) = 0.$$

Solution:

The given equation is $ay^4 + bxy^3 + cx^2y^2 - dxy^3 + ex^4 = 0$.

If two of the lines represented by (1) at right angles, their equation must be of the form $x^2 + pxy - y^2 = 0$ as the sum of the coefficients of x^2 and y^2 is zero. Hence let

$$ay^4 + bxy^3 + cx^2y^2 + dyx^3 + ex^4 \equiv (x^2 + pxy - y^2)(ex^2 + qxy - ay^2)$$

As the coefficient of x^4 on both sides is same, other coefficients with also be equal.

Now compare different coefficients.

Comparing xy^3, we have $-pa - q = b$. ...(2)

Comparing x^2y^2, we have $-a + pq - e = c$. ...(3)

Comparing x^3y, we have $q + eq = d$

Now eliminate p and q from (2), (3) and (4).

From (2) and (4) $p = \frac{b + d}{e - a}$

and $q = -\dfrac{da+eb}{e-a}$.

Substituting these values in (3), we get

$$-a - \frac{b+d}{e-a}.\frac{da+eb}{e-a} - e = c$$

$\Rightarrow$ $(a + e + c)(e - a)^3 + (b + d)(ad + be) = 0.$

Example 46(c):

Hence find the locus of the orthocentre of a triangle of which two sides are given in position and whose third side goes through a fixed point.

Solution:

Let the line $\lambda x + my = 1$ passes through some fixed point $P \equiv (\alpha, \beta)$. Then $l\alpha + m\beta = 1$. ...(1)

If (x', y') be the orthocentre,

$$\frac{x'}{l} = \frac{y'}{m} = \frac{a+b}{am^2 - 2hlm + bl^2} + \frac{x'+y'}{l+m} \qquad ...(2)$$

As $\dfrac{x'}{l} = \dfrac{y'}{m} = \dfrac{\alpha x'}{\alpha l} = \dfrac{\beta y'}{\beta m} = \dfrac{\alpha x'+\beta y'}{\alpha l + \beta m} = \alpha x' + \beta y'$ [as $al + \beta m = 1$ by (1)]

$\therefore$ $\dfrac{x'}{l} = \dfrac{y'}{m} = \alpha x' + \beta y'$.

$\therefore$ $l = \dfrac{x'}{\alpha x'+\beta y'}, m = \dfrac{y'}{\alpha x'+\beta y'}$.

Putting these values in (2), we get

$$\frac{a+b}{a\left(\dfrac{y'}{\alpha x'+\beta y'}\right)^2 - 2h\dfrac{x'}{\alpha x'+\beta y'}.\dfrac{y'}{\alpha x'+\beta y'} + b\left(\dfrac{x'}{\alpha x'+\beta y'}\right)^2}$$

$$= \frac{x'+y'}{\dfrac{x'}{\alpha x'+\beta y'} + \dfrac{y'}{\alpha x'+\beta y'}}$$

$\Rightarrow$ $\dfrac{(a+b)(\alpha x'+\beta y')^2}{ay'^2 - 2hx'y' + bx'^2} = \dfrac{(x'+y')(\alpha x'+\beta y')}{(x'+y')}$.

Simplifying and generalising for x' and y', we get the required locus as $(a + b)(\alpha x + \beta y) = ay^2 - 2hxy + bx^2$.

Example 47:

Prove that one of the lines represented by the equation $ax^3 + bx^2y + cxy^2 + dy^3 = 0$, will bisect the angle between the other two if $(3a + c)^2 (bc + 2cd - 3ad) = (b + 3d)^2 (bc + 2ab - 3ad)$.

Solution:

The given line is $ax^3 + bx^2y + cxy^2 + dy^3 = 0$ ...(1)

This being homogeneous cubic equation represents three straight lines each passing through the origin.

Let these lines be $y - m_1x = 0,$ ...(2)

$y - m_2x = 0,$ ...(3)

and $y - m_3x = 0.$ ...(4)

Let the equation (4) be the bisector of the angle between (2) and (3). Then, if the angles that the lines (2), (3) and (4) make with x-axis be θ_1, θ_2 and θ. clearly $2\theta = \theta_1 + \theta_2$ or $\tan 2\theta = \tan(\theta_1 + \theta_2)$

$$\Rightarrow \quad \frac{2\tan\theta}{1-\tan^2\theta} = \frac{\tan\theta_1 + \tan\theta_2}{1-\tan\theta_1\tan\theta_2} = \frac{3m_3}{1-m_3^2} = \frac{m_1+m_2}{1-m_1m_2} \quad ...(5)$$

Again equation no. (1) is same as represented by (2), (3) and (4).

Hence $ax^3 + bx^2y + cxy^2 + dy^3 \equiv (y - m_2x)(y - m_3x)$.

Comparing the coefficients, we get

$$\frac{1}{d} = \frac{-(m_1+m_2+m_3)}{c} = \frac{m_1m_2 + m_2m_3 + m_3m_1}{b} = \frac{-m_1m_2m_3}{a}$$

$$\Rightarrow \quad m_1 + m_2 + m_3 = -\frac{c}{d}, \quad ...(6)$$

$$m_1m_2 + m_2m_3 + m_3m_1 = \frac{b}{d} \quad ...(7)$$

and $$m_1m_2m_3 = -\frac{a}{d} \quad ...(8)$$

Putting the value of $m_1 + m_2$ from (6) in (5), we get

$$\frac{2m_3}{1-m_3^2} = \frac{-\frac{c}{d} - m_3}{1-m_1m_2} = \frac{-c-dm_3}{d-dm_1m_2}$$

$$\Rightarrow \quad 2m_3d - 2dm_1m_2m_3 = -c - dm_3 = cm_3^2 + dm_3^2$$

$$\Rightarrow \quad dm_3^2 + cm_3^2 - c + 2d\left(-\frac{a}{d}\right) = 0 \quad ...(9)$$

$$\left\{\therefore \; m_1m_2m_3 = -\frac{a}{d} \text{ by } (8)\right\}$$

Now dividing (1) by x^2, we get, $a+b\frac{y}{x}+c\left(\frac{y}{x}\right)^2+d\left(\frac{y}{x}\right)^3=0.$

As $y = m_3$ and (10), on substraction, we get m_3 (b + 3d) = – (3a + c)

So $$a + bm_3 + cm_3^2 + dm_3^2 = 0\,. \qquad ...(10)$$

From (9) and (10), on substraction, we get m_3 (b + 3d) = – (3a + c)

or $$m_3 = \frac{3a+c}{b+3d}.$$

Substituting this value in (10) and simplifying, we get the required result as $(3a + c)^2 (bc + 2cd - 3ad) = (b + 3d)^2 (bc + 2ab - 3ad)$.

Example 48:

Show that the orthocentre of the triangle formed by the straight lines $ax^2 + 2hxy + by^2 = 0$ and $lx + my + l$ is a point (x', y') such that

$$\frac{x'}{l}=\frac{y'}{m}=\frac{a+b}{am^2-2hlm+bl^2}.$$

Solution:

Let the equation $ax^2 + 2hxy + by^2 = 0$ represent OB and OA respectively as

$$y = m_1x \qquad ...(1)$$

and $$y = m_2x \qquad ...(2)$$

Then $m_1 + m_2 = -\frac{2h}{b}$ and $m_1 m_2 = \frac{a}{b}$

The line AB is given as $lx + my = 1$. ...(3)

Solving (1) and (3), we get $x = \frac{1}{l+mm_1}\, y = \frac{m_1}{l+mm_1}$

Hence the co-ordinates of B are $\left(\frac{1}{l+mm_1}\;\frac{m_1}{l+mm_1}\right)$.

The equation of the line perpendicular to OB i.e., (2), passing to OB i.e., (2), passing through opposite vertex A, will be

$$y-\frac{1}{l+mm_1}=-\frac{1}{m_2}\,x-\left(\frac{1}{l+mm_1}\right)$$

or $(l + mm_1)\, x\, m_2\, (l + mm_1)\, y - (m_1m_2 + 1) = 0.$

As the line passes through the orthocentre (x', y'), so it will satisfy it. Hence,

$$(l + mm_1)\, x' + m_2\, (l + mm_1)\, y' - (m_1 m_2 + 1) = 0. \quad \text{...(4)}$$

Again, the line perpendicular to AB passing through the vertex O (0, 0) will be $mx - l\, y = 0$.

As it also passes through the orthocentre (x', y'), hence

$$mx' - ly' = 0. \quad \text{...(5)}$$

To solve (4) and (5), putting value of y' from (5) in (4); we get

$$\left(l + mm_1\right) x' + m_2 \left(l + mm_1\right) \frac{mx'}{l} - \left(m_1\, m_2 + 1\right) = 0.$$

$$\Rightarrow \quad \frac{x'}{l}\left[l^2 + l\, mm_1\, l + mm_2\, l + m^2\, m_1\, m_2\right] = 1 + m_1\, m_2$$

$$\Rightarrow \quad \frac{x'}{l}\left[l^2 + l\, m\left(m_1 + m_2\right) + m_1\, m_2\right] = 1 + m_1\, m_2\,.$$

Putting the values of $m_1 + m_2$ and $m_1\, m_2$ as $-\frac{2h}{b}$ and $\frac{a}{b}$ respectively, we get

$$= \frac{x'}{l} \cdot \left[l^2 - l\, m \frac{2h}{b} + m^2\, \frac{a}{b}\right] = 1 + \frac{a}{b} \text{ or } \frac{x''}{l} = \frac{a + b}{l^2 b - 2hlm + am^2}$$

Similarly putting the value of x' from (5) in (4), we get

$$\frac{y'}{l} = \frac{a + b}{l^2 b - 2hlm + am^2}.$$

Hence $\quad \dfrac{x'}{l} = \dfrac{y'}{m} = \dfrac{a + b}{l^2 b - 2hlm + am^2}.$

Example 49:

Transform to axes inclined at 45° to the original axes the equations (1) $x^2 - y^2 = a^2$, (2) $17x^2 - 16xy + 17y^2 = 225$, and (3) $y^4 + x^4 + 6x^2 y^2 = 2$.

Solution:

(i) Given equation is $x^2 = y^2 = a^2$. The axes are turned to an angle of 45°, hence putting (x' cos 45° – y' sin 45°) for x and

(x' sin 45° + y' cos 15°) for y, we get

$$\Rightarrow \quad \frac{1}{2}\left[(x' - y')^2 - (x' + y')^2\right] = a^2 \;\therefore\; \cos 45^\circ = \sin 45^\circ = \frac{1}{\sqrt{2}}$$

$\Rightarrow \qquad 2x'y' + a^2 = 0.$

(ii) If the curve is $17x^2 - 16xy + 17y^2 = 225$, by the same substitution, we get

$$17\,(x' \cos 45 - y' \sin 45)^2 - 16\,(x' \cos 45 - y' \sin 45)$$
$$\times\,(x' \sin 45 + y' \cos 45) + 17\,(x' \sin 45 + y' \cos 45)^2 = 225$$

$\Rightarrow \frac{1}{2}\left[18x^2 + 50y^2\right] = 225$ or $9x^2 + 25y^2 = 225.$

(iii) Curve is $y^4 + x^4 + 6x^2y^2 = 2.$

By the same substitution, we get

$$(x' \cos 45 + y' \sin 45)^4 + (x' \sin 45 - y' \cos 45)^4$$
$$+ 6\,(x' \cos 45 - y' \sin 45)^2\,(x' \sin 45 + y' \cos 45)^2 = 2$$

$\therefore x'^4 + y'^4 = 1$ (on simplification).

Example 50:

Transform to axes inclined at an angle α to the original axes the equations (1) $x^2 + y^2 = r^2$, and (2) $x^2 + 2xy \tan 2\alpha - y^2 = a^2$.

Solution:

The given equation is $x^2 + y^2 = r^2$.

(i) The axes are turned to an angle α. Substituting the values of x and y in x', and α, we get, $(x' \cos \alpha - y' \sin \alpha)^2 + (x \sin + y' \cos \alpha)^2 = r^2$.

Simplifying, we get $x'^2 + y'^2 = r^2$.

(ii) $x^2 + 2xy \tan 2\alpha - y^2 = a^2$. By the same substitution, we get

$$(x' \sin \alpha - y' \sin \alpha)^2 + 2\,(x' \cos \alpha - y' \sin \alpha)$$
$$\times\,(x' \sin \alpha + y' \cos \alpha) \tan 2\alpha - (x' \sin \alpha + y' \cos \alpha)^2 = a^2$$

$$\Rightarrow x'^2 (\cos^2 \alpha - \sin^2 \alpha) - y'^2 (\cos^2 \alpha - \sin^2 \alpha) - 2x'y' \sin \alpha \cos \alpha$$
$$+ [2 \sin \alpha \cos \alpha\,(x'^2 - y'^2) + x'y'\,(\cos^2 \alpha - \sin^2 \alpha] \tan 2\alpha = a^2$$

$$\Rightarrow x'^2 \cos 2\alpha - y'^2 \cos 2\alpha - x'y' \sin 2\alpha$$
$$= \left[\left(x'^2 - y'^2\right) \sin 2\alpha + x'y' \cos 2\alpha\right] \frac{\sin 2\alpha}{\cos 2\alpha} = a^2$$

$$\Rightarrow (x'^2 - y'^2) \cos^2 2\alpha - x'y' \sin 2\alpha + (x'^2 - y'^2) \sin^2 2\alpha$$
$$+ x'y' \sin 2\alpha = \alpha^2 \cos 2\alpha.$$

$$\Rightarrow x'^2 - y'^2 = a^2 \cos 2\alpha.$$

Example 51:

Transform the equation $y^2 + 4y \cot \alpha - 4x = 0$ from rectangular axes to oblique axes meeting at an angle α, the axis of x being kept the same.

Solution:

The given equation is $y + 4y \cot \alpha - 4x = 0$. ...(1)

By hypothesis, $\omega = 90°$, $\omega' = \alpha$, $\theta = 0$.

For transformation $x \sin 90° = x' \sin (90 - 0) + y' \sin (90 - \alpha - 0)$

$\Rightarrow \quad x = x' + y' \cos \alpha$ ...(2)

and $\quad y \sin 90° = x' \sin 0 + y' \sin (\alpha + 0)$

$\Rightarrow \quad y = y' \sin \alpha.$ ...(3)

Putting the values from (2) and (3) in (1), we get

$$(y' \sin \alpha)^2 + 4y' \sin \alpha . \frac{\cos \alpha}{\sin \alpha} - 4 . (x' + y' \cos \alpha) = 0$$

$\Rightarrow \quad y'^2 \sin^2 \alpha + 4y' \cos \alpha - 4x' - 4y' \cos \alpha = 0$

$\Rightarrow \quad y'^2 \sin^2 \alpha + 4x'$

or $\quad y'^2 = 4x' \operatorname{cosec}^2 \alpha.$

Example 52:

Transform the equation $2x^2 + 3\sqrt{3xy} + 3y^2 = 2$ from axes inclined at 30° to rectangular axes, the axis of remaining unchanged.

Solution:

The given equation is $2x^2 + 3\sqrt{3xy} + 3y^2 = 0$. ...(1)

By hypothesis, $\omega = 30' = 90°$, $\theta = 0$, hence we have

$x \sin 30 = \omega' = 90°$, $\theta = 0$, hence we have

$\Rightarrow \quad \frac{1}{2}.x = x'.\frac{1}{2} - \frac{\sqrt{3}}{2}.t'$ or $x = x' - \sqrt{3y'}$ ...(2)

and $\quad y \sin 30 = x' \sin = 0 + y' \sin (90 + 0)$

$\Rightarrow \quad$ Substituting from (2) and (3) in (1), we get

$2 (x' - \sqrt{3y'})^2 + 3\sqrt{3} (x' - \sqrt{3y}') + 3 (2y)^2 = 2$

$\Rightarrow \quad x'^2 + \sqrt{3.x'y'} = 1$ (on simplification).

Example 53:

Transform the equation $x^2 + xy + y^2 = 8$ from axes inclined at 60° to axes bisecting the angles between the original axes.

Solution:

The origin axes are inclined at angle of 60°.

Bisectors of the angles will make angles of 30° and 120° from original axis of x. Hence here $\omega = 30°$, $\omega' = 90°$, $\theta = 30°$.

For transformation,

$$x \sin 60° = x' \sin (60° - 30°) + y' \sin (60° - 90° - 30°)$$

$$\Rightarrow \quad x.\frac{\sqrt{3}}{2} = x'.\frac{1}{2} - y'\frac{\sqrt{3}}{2} \text{ or } x = \frac{x'}{\sqrt{3}} - y'. \quad ...(1)$$

And $y \sin 60° = x' \sin 30 + y' \sin (90 + 30)$

$$\Rightarrow \quad y.\frac{\sqrt{3}}{2} = x'.\frac{1}{2} + y'\frac{\sqrt{3}}{2} \text{ or } y = \frac{x'}{\sqrt{3}} x' + y' \quad ...(2)$$

The original equation is given as $x^2 + xy + y^2 = 8$.

Substituting the values of x and y from (1) and (2), we get

$$\left(\frac{x'}{\sqrt{3}} - y'\right)^2 + \left(\frac{x'}{\sqrt{3}} - y'\right)\left(\frac{x'}{\sqrt{3}} + y'\right) + \left(\frac{x'}{\sqrt{3}} + y'\right)^2 = 8$$

$$\Rightarrow \quad x'^2 + y'^2 = 8 \text{ (on simplification).}$$

Example 54:

The equation to a straight line referred to axes inclined at 30° to one another is $y = 2x + 1$. Find its equation referred to axes inclined at 45°, the origin and axis of x being unchanged.

Solution:

The given equation is $y = 2x + 1$. ...(1)

by hypothesis, $w = 30°$, $w' = 45°$ and $q = 0$. So we have the substitution $x \sin 30° = x' \sin (30 - 0) + y' \sin (30 - 45 - 0)$

$$\Rightarrow \quad \frac{1}{2}x = x'\frac{1}{2} - y'\frac{\sqrt{3}-1}{2\sqrt{2}} \text{ or } x = x' - \frac{\sqrt{3}.-1}{\sqrt{2}} y' \quad ...(2)$$

and $\quad y \sin 30° = x' \sin 0 + y' \sin 45$

$$\Rightarrow \quad y.\frac{1}{2} = y'.\frac{1}{\sqrt{2}} \text{ or } y = \sqrt{2}.y' \quad ...(3)$$

Putting the values from (2) and (3) in (1), we get

$$\sqrt{2}.y' = 2.\left(x' - \frac{\sqrt{3}-1}{\sqrt{2}}y'\right) + 1$$

$$2x' - \sqrt{6y'} + 1 = 0.$$

Example 55:

If x and y be the coordinates of a point referred to a system of oblique axes, and x' and y' be its coordinates referred to another system of oblique axes with the same origin, and if the formulae of transformation be

$$x = mx' + ny'$$

and $$y = m'x' + n'y',$$

prove that $$\frac{m'^2 + m'^2 - 1}{n'^2 + n'^2 - 1} = \frac{mm'}{nn'}.$$

Solution:

Let the original axes be inclined at an angle of ω, the origin being O. If there is any point P whose co-ordinates with respect to these axes be (x, y); then

$$OP^2 = x^2 + y^2 + 2xy \cos \omega. \qquad ...(1)$$

If the axes are changed and the angle between the new axes be ω', origin remaining the same, let the co-ordinates of P with respect to new axes be (x', y')' then

$$OP^2 = x'^2 + y'^2 + 2x'y, \cos \omega'. \qquad ...(2)$$

By (1) and (2), we get

$$x^2 + y^2 + 2xy \cos \omega = x'^2 + y'^2 + 2x'y' \cos \omega'. \qquad ..(3)$$

As the transformation are given as,

$$x = mx' + ny' \text{ and } y = m'x' + n'y'.$$

Putting these values in (3), we get

$$(mx' + ny)^2 + (m'x' + n'y')^2 + 2\,(mx' + ny')\,(m'x' + n'y') \cos \omega.$$

$$= x'^2 + y'^2 + 2x'y' \cos \omega. \qquad ...(4)$$

Comparing the coefficients of x'^2 on both sides of (4), we have

$$m^2 + m'^2 = 2mm' \cos \omega = 1.$$

$$\Rightarrow \quad m^2 + m'^2 - 1 = -2nn' \cos \omega \qquad ...(6)$$

Dividing (6) by (5), we get

Comparing the coefficients of y'^2 in (4), we have

$$n^2 + n'^2 = 2mm' \cos \omega = 1.$$

$$\Rightarrow \quad n^2 + n'^2 - 1 = -2nn' \cos \omega \qquad ...(6)$$

Dividing (6) by (5), we get

$$\frac{n^2 + n'^2 - 1}{m^2 + m'^2 - 1} = \frac{nn'}{mm'},$$

$$\Rightarrow \quad \frac{m^2 + m'^2 - 1}{mm'} = \frac{n^2 + n'^2 - 1}{nn'}.$$

Example 56:

Transform to parallel axes through the point (1, –2) the equations (1) $y^2 - 4x + 4y + 8 = 0$ and (2) $2x^2 + y^2 - 4x = 0$.

Solution:

(i) The equation is $y^2 - 4x + 4y + 8 = 0$. The origin is transferred to (1, –2). So the new equation will be

$$(y' - 2)^2 - 4(x' + 1) + 4(y' - 2) + 8 = 0$$

$$\Rightarrow \quad y'^2 - 4y' + 4 - 4x' - 4 + 4y' - 8 + 8 = 0$$

$$\Rightarrow \quad y'^2 = 4x'.$$

(ii) The equation is $2x^2 + y^2 - 4x + 4y = 0$.

Transferring the origin to (1, –2), we get

$$2(x' + 1)^2 + (y' - 2)^2 - 4(x' + 1) + 4(y' - 2) = 0$$

$$\Rightarrow \quad 2x'^2 + 4x' + 2 + y'^2 - 4y' + 4 - 4 + 4y' - 8 = 0$$

$$\Rightarrow \quad 2x'^2 + y'^2 = 6.$$

Example 57:

What does the equation $(x - a)^2 + (y - b)^2 = c^2$, become when it is transferred to parallel axes through (1) the point (a – c, b), (2) the point (a, b, – c)?

Solution:

The curve is given as $(x - a)^2 + (y - b)^2 = c^2$.

(i) Transferring the origin to the point (a – c, b), we get

$$[x' + (a - c) - a]^2 + [y' + b - b)^2] = c^2$$

$$\Rightarrow \quad x'^2 + c^2 - 2x'c + y'^2 = c^2 \text{ or } x'^2 + y'^2 = 2cx'.$$

(ii) Transferring the origin to the point (a, b – c), we get

$$[x' + a - a]^2 + [y' + b - c - b]^2 = c^2$$

$\Rightarrow \quad x'^2 + y'^2 + c^2 - 2y'c = c^2$

or $\quad x'^2 + y'^2 = 2y'c.$

Example 58:

What does the equation $(a - b)(x^2 + y^2) - 2abx = 0$, become if the origin be moved to the point $(ab/a - b, 0)$?

Solution:

The equation is $(a - b)(x^2 + y^2) - 2abx = 0$.

The origin is transferred to $(ab/a - b, 0)$; hence the equation becomes

$$(a-b)\left[\left\{x'+\frac{ab}{a-b}\right\}^2 + \{y'+0\}^2\right] - 2ab.\left[x'+\frac{ab}{a-b}\right] = 0$$

$$\Rightarrow \quad (a-b)\left(x'^2+y'^2\right)+\frac{a^2\ b^2}{a-b}+\text{lab}.x'-2abx'-\frac{2a^2\ b^2}{a-b}=0$$

$$\Rightarrow \quad (a-b)\left(x'^2+y'^2\right)+\frac{a^2\ b^2}{a-b} \text{ or } (a-b)^2\left(x'^2+y'^2\right)=a^2\ b^2.$$

Example 59:

Find the angle through which the axes may be turned so that the equation $Ax + By + C = 0$, may be reduced to the form x = constant and determine the value of this constant.

Solution:

Let the required angle be α. Then turning the axes to the angle θ, origin being the same, the given equation $Ax + By + C = 0$ changes to $A(x' \cos \alpha - y' \sin \alpha) + B(x' \sin \alpha + y' \cos \alpha) + C = 0$.

or $x'(A \cos \alpha + B \sin \alpha) - y'(A \sin \alpha - B \cos \alpha) + C = 0$ ...(1)

If the equation takes the form x = constant, then the coefficient of y' must be zero; hence $A \sin \alpha - B \cos \alpha = 0$ or $A \sin \alpha = B \cos \alpha$ or $\tan \alpha = B/A$.

Hence the axes should be changed to an angle $\tan^{-1}(B/A)$.

Putting this value in (1), we get

$$x'\left[A.\frac{A}{\sqrt{(A^2+B^2)}}+B.\frac{B}{\sqrt{(A^2+B^2)}}\right]+C=0,$$

$\therefore$ if $\tan\alpha = \dfrac{B}{A}$ then $\cos\alpha = \dfrac{A}{\sqrt{(A^2+B^2)}}$ and $\sin\alpha = \dfrac{B}{\sqrt{(A^2+B^2)}}$

$$\Rightarrow \qquad x = -\frac{C\sqrt{(A^2+B^2)}}{A^2+B^2} = -\frac{C}{\sqrt{(A^2+B^2)}}$$

So required constant is $-\dfrac{C}{\sqrt{(A^2+B^2)}}$.

Example 60:

If the axes be turned through an angle tan^{-1} 2, what does the equation $4xy - 3x^2 = a^2$ become ?

Solution:

The given equation is $4xy - 3x^2 = a^2$ and the angle a = $\tan^{-1} 2$ or $\tan\alpha = 2$; hence $\sin\alpha = 2/\sqrt{5}$ and $\cos\alpha = 1/\sqrt{5}$.

Substituting the appropriate values of x and y, we get

$$4(x'\cos\alpha - y'\sin\alpha)(x'\sin\alpha + y'\cos\alpha) - 3(x'\cos\alpha - y'\sin\alpha)^2 = a^2$$

$$\Rightarrow \qquad 4\left[\frac{1}{\sqrt{5}}x' - \frac{1}{\sqrt{5}}y'\right]\left[\frac{2}{\sqrt{5}}x' + \frac{1}{\sqrt{5}}y'\right] - 3\left[\frac{1}{\sqrt{5}}x' - \frac{2}{\sqrt{5}}y'\right]^2 = a^2$$

$\Rightarrow \qquad x'^2 - 4y'^2 = a^2$ (on simplification).

Example 61:

By transferring so parallel axes through a properly chosen point (h, k), prove that the equation $12x^2 - 10xy + 2y^2 + 11x - 5y + 2 = 0$, can be reduced to one containing only terms of the second degree.

Solution:

The given equation is

$$12x^2 - 10xy + 2y^2 + 11x - 5y + 2 = 0.$$

Let the origin be transferred to (h, k) axes being parallel to the previous axes; then the equation becomes

$$12(x'+h)^2 - 10(x'+h)(y'+k)^2 + 11(x'+h) - 5(y'+k) + 2 = 0$$

$$\Rightarrow 12x'^2 + 12h^2 + 24x'h - 10x'y' - 10x'k - 10y'h - 10hk + 2y'^2 + 2k^2 + 4y'k + 11x' + 11h - 5y' - 5k + 2 = 0 \qquad ...(1)$$

$$\Rightarrow 12x'^2 + 2y'^2 - 10x'y' + x'(24h - 10k + 11) + y'(-10h + 4k - 5)$$
$$+ 12h^2 - 10hk + 2k^2 + 11h - 5k + 2 = 0.$$

If this equation contains the terms of x^2 and y^2 and constant terms only, then the coefficients of x' and y' must be zero.

So $\qquad 24h - 10k + 11 = 0 \qquad$ ...(2)

and $\qquad (-10h + 5k - 5) = 0.$

Solving (2) and (3), we get $h = \frac{3}{2}$ and $k = -\frac{5}{2}$

Hence the required point is $\left(-\frac{3}{2}, -\frac{5}{2}\right)$,

If we substitute the values in (1), the equation reduces to

$$12x^2 - 10x'y' - 2y'^2 = 0.$$